T0328418

INTEGRATED OPERATION OF HYDROPOWER STATIONS AND RESERVOIRS

EXPLOITATION DES CENTRALES HYDROÉLECTRIQUES ET DES RÉSERVOIRS

INTERNATIONAL COMMISSION ON LARGE DAMS
COMMISSION INTERNATIONALE DES GRANDS BARRAGES
61, avenue Kléber, 75116 Paris
Téléphone : (33-1) 47 04 17 80
http://www.icold-cigb.org

Cover/Couverture : Some hydropower projects in the world / *Quelques projets hydroélectriques dans le monde*

CRC Press/Balkema is an imprint of the Taylor & Francis Group, an informa business
© 2023 ICOLD/CIGB, Paris, France

Typeset by CodeMantra
Published by: CRC Press/Balkema
Schipholweg 107C, 2316 XC Leiden, The Netherlands
e-mail: enquiries@taylorandfrancis.com
www.routledge.com – www.taylorandfrancis.com

Original text in English
French translation by Thierry Avril (France), Patrice Droz (Switzerland),
R Kamanga (Zambia), Louise Remilliard (Canada).
Layout by Nathalie Schauner

Texte original en anglais
Traduction en français par Thierry Avril (France), Patrice Droz (Suisse),
R Kamanga (Zambie), Louise Remilliard (Canada).
Mise en page par Nathalie Schauner

ISBN: 978-0-367-77005-1 (Pbk)
ISBN: 978-1-003-16934-5 (eBook)

COMMITTEE ON INTEGRATED OPERATION OF HYDROPOWER STATIONS AND RESERVOIRS

COMITÉ DE L'EXPLOITATION INTÉGRÉE DES CENTRALES HYDROÉLECTRIQUES ET DES RÉSERVOIRS

Chairman/Président

China / Chine	CHEN Guoqing

Members/Membres

Brazil / Brésil	Luciano VARELLA
Canada	Jean MATTE
China / Chine	LI Hui (co-opted)
France	Thierry AVRIL
Germany / Allemagne	Christian HEITEFUSS
Iran	Abdolrahim SALAVITABAR
Japan / Japon	Hiroyuku DOI
Korea / Korée	Kee-uk CHA
Nigeria	Johnson ADEWUMI
Norway / Norvège	Bjørn HONNINGSVÅG
Russia / Russie	Alexander GRIGOREV
South Africa / Afrique du Sud	Leon FURSTENBURG
Switzerland /Suisse	Patrice DROZ
Turkey / Turquie	Ertan DEMIRBAS
United States / États Unis	David BOWLING
Zambia / Zambie	Victor MUNDENDE

SOMMAIRE	CONTENTS

TABLE DES MATIÈRES

TABLE OF CONTENTS

FIGURES & TABLEAUX

FIGURES

FIGURES & TABLES

FIGURES

TABLEAUX

TABLES

PRÉAMBULE

Sur la terre l'eau est nécessaire à la vie. L'eau est donc un élément essentiel à tous les êtres vivants et particulièrement à l'Homme. Il n'est donc pas étonnant que nos ancêtres aient développé depuis des millénaires des ouvrages pour subvenir à leurs besoins en eau potable, ou encore en matière d'agriculture. Ainsi en Chine et dans les régions babyloniennes, des réservoirs ont été construits depuis l'antiquité. Ces besoins sont toujours vitaux aujourd'hui. Les aménagements hydroélectriques donnent accès à l'eau potable, à l'irrigation des terres, à la production d'énergie mécanique et électrique, ils contribuent au transport des biens et des personnes, etc.

Le développement de l'hydroélectricité a commencé autour de 1870 pour l'industrie du papier et avec les inventions des turbines Francis (1868) et Pelton (1879). La première centrale hydroélectrique a été construite en 1882 en associant une turbine et une dynamo de Gramme.

Quelques décennies plus tard, la tendance vers un développement hydroélectrique couvrant un bassin entier apparu d'abord au Japon. Dans les années 1930, la cascade d'aménagements multi-objectifs de la vallée du Tennessee était initiée. Au même moment, l'URSS lançait son programme d'aménagement de la Volga. A présent, de nombreux aménagements en cascade ont été développés tels que ceux de la rivière Columbia, de la rivière Tennessee, de la rivière La Grande en Amérique du nord, des rivières Paraná et Orénoque en Amérique du sud, des rivières Ariège, Rhône, Douro et Volga Kama en Europe, des rivières Ieinisseï, Kiso et grand Karun en Asie, ainsi que le Nil, le Niger et le Zambèze en Afrique. En Chine, le développement du haut Yang-Tsé et de ses affluents est en cours, il sera dans un proche futur le plus grand complexe hydroélectrique du monde.

Pour faciliter le développement d'aménagements intégrés et en avoir une meilleure compréhension, Le CIGB décida de créer un comité traitant de l'exploitation des aménagements hydroélectriques et des réservoirs en 2011. Les activités de ce comité ont débuté lors de la réunion de Kyoto en juin 2012.

Les objectifs étaient les suivants :

- Exploitation sûre des aménagements hydroélectriques,

- Exploitation intégrée des aménagements hydroélectriques comprenant des réservoirs à buts multiples,

- Exploitation optimale des réservoirs et aménagements hydroélectriques en cascade,

- Publication de recommandations pour la gestion, l'exploitation et la maintenance d'aménagements hydroélectriques. Dans le but d'améliorer leur sécurité, leur efficacité et leur qualité d'exploitation.

Le comité publie ce bulletin pour servir de référence aux lecteurs dans le domaine de l'hydroélectricité et des domaines liés aux autres usages de l'eau. Il donne un aperçu des principes fonctionnels de base relatifs aux aménagements hydrauliques enchaînés, établis à partir d'études de cas produits par les différents pays membres participant au comité. Ce bulletin a été construit à partir d'une revue de l'ensemble des aspects présentés par ces études de cas.

J'ai apprécié le grand effort du précédent Président, le Dr Cao Guangjing, ainsi que la sincère coopération et la participation enthousiaste de chacun des 16 membres du Comité, qui ont partagé leur expérience et leur temps pour élaborer ce bulletin.

J'espère que ce bulletin servira à donner une vision holistique des différents aspects liés aux fonctions des aménagements hydroélectriques en cascade.

CHEN Guoqing
Président,
Comité pour l'Exploitation Intégrée des Réservoirs et Aménagements Hydroélectriques

FOREWORD

Life on the earth needs water. Water is an essential need for life and particularly human beings. It is not astonishing that our ancestors developed works for supplying drinkable water and then for agricultural purposes in thousands of years. In China and in Babylonian regions, reservoirs have been built from the antiquity along with the development of agriculture. This need is still vital today. Hydroelectric schemes give access to drinkable water, lands irrigation, production of mechanical and electric energy, participation to transport of people and goods and so on.

Hydropower development started in around the 1870s in the paper industry with the invention of Francis (1868) and Pelton (1879) turbines. The first hydropower station was built in France in 1882 by combining a turbine and a Gramme dynamo.

Decades later, the trend for cascade hydropower developments in a whole basin first appeared in Japan. Then in the 1930s, multi-objective comprehensive cascade development of the Tennessee Valley was proposed, and implementation started. At the same time, the USSR started to plan and implement the Volga River development. Currently, there are many well-known comprehensive cascade developments on larger rivers, such as the Columbia River, the Tennessee River and the La Grande River in North America, the Parana River and the Orinoco River in South America, the Ariège River, the Rhone River, the Douro River and the Volga-Kama river in Europe, the Yenisei River, the Kiso River and the Great Karun River in Asia, as well as the Nile River, the Niger River and the Zambezi River in Africa. In China, the comprehensive development in the upper stream of Yangtze and its tributaries is ongoing and the world's largest hydropower system will be formed there in the near future.

In order to facilitate the development of comprehensive cascades and have a better understanding of them, ICOLD decided to organize the Committee on Integrated Operation of Hydropower Stations and Reservoirs and appointed the committee in 2011. Committee activities started from the Kyoto Conference in June 2012.

The terms of reference for the Committee were:

- Safe operation and management of hydropower stations.

- Integrated operation of hydropower stations with multi-objective-oriented reservoirs.

- The optimal operation of basin reservoirs and hydropower stations for cascade developments.

- Publishing of guidelines for management, operation and maintenance of hydropower stations in order to provide a reference and basis for improving their safety, efficiency and management level.

The Committee publishes this bulletin as a reference for readers in hydropower and related fields. It gives an overview of the main functional aspects relating to cascade hydropower stations and typical case studies in member countries. It was formed by reviewing of all the related aspects proposed and case studies provided by committee members.

I appreciate the great effort made by former chairman Dr. Cao Guangjing as well as sincere cooperation and enthusiastic participation of all 16 committee members who shared their experience and time for the creation of this bulletin.

I hope that this bulletin serves to give a holistic vision of the various aspects related to the functions and operations of cascade hydropower stations.

CHEN GUOQING
Chairman,
Committee on Integrated Operation of Hydropower Stations and Reservoirs

1. INTRODUCTION

Les barrages servent à accumuler de l'eau pour des finalités diverses. Ainsi l'accumulation d'eau peut servir à élever le niveau d'eau dans une rivière afin d'en faciliter la navigation sur un certain tronçon, ou créer une charge hydraulique suffisante pour générer de l'énergie. Le potentiel d'accumulation peut également servir à retenir une partie de l'eau lors d'épisodes de crue, celle-ci pouvant être utilisée par la suite pour l'alimentation en eau potable, en eau industrielle, ou pour l'irrigation. Enfin elle peut également être utilisée pour des enjeux écologiques ou pour la production d'énergie. Dans certains cas, le niveau d'eau dans la retenue est maintenu minimal afin de garantir un effet protecteur contre les crues à l'aval de l'ouvrage. Ces finalités peuvent être souvent contradictoires. Enfin les réservoirs artificiels peuvent offrir des intérêts touristiques et de loisirs indéniables. Ainsi, les barrages doivent être conçus et exploités de sorte à assurer qu'ils remplissent leurs objectifs de façon économique et durable.

Les barrages sont généralement construits sur des rivières. Ainsi, la structure peut présenter une discontinuité le long du système du cours d'eau naturel, ce qui peut affecter particulièrement le transport des sédiments, la migration piscicole ou la reproduction des espèces. Les variations de débit en aval ainsi que les nouvelles berges créées autour du réservoir impactent l'environnement riverain : ces considérations environnementales induisent ainsi souvent des contraintes d'exploitation.

Les barrages, en tant que structures construites dans les rivières, doivent être de plus capables de supporter en toute sécurité, pendant de longues périodes, les effets des crues et des sédiments.

Très souvent, les barrages sont conçus et exploités non pas dans un seul but, mais pour plusieurs objectifs. Les différents objectifs peuvent engendrer des conceptions ou des modes d'exploitations conflictuels, si bien qu'un compromis économique est souvent nécessaire. Cette situation est souvent compliquée par le fait que plusieurs barrages peuvent être construits en aval l'un de l'autre formant une cascade d'aménagements. Dans de nombreux pays, la complexité est augmentée par des transferts d'eau entre bassins versant, si bien que l'exploitation intégrée doit s'étendre à l'ensemble des aménagements et des bassins. De nombreuses approches de ces systèmes complexes ont été faites depuis des années pour simuler de façon efficace, optimiser et contrôler leur exploitation.

En raison du large éventail des usages de l'eau nécessitant une accumulation, une planification et une exploitation intégrée de la ressource en eau est nécessaire. Ces pratiques sont bien développées dans plusieurs pays de par le monde.

1.1. LE RÔLE DES BARRAGES

Partout dans le monde, les barrages ont été conçus, construits et exploités comme faisant partie intégrale de la gestion de la ressource en eau et du développement. Historiquement, la plupart des barrages étaient construits afin de suppléer, en premier lieu, aux besoins de l'irrigation (Mésopotamie et en Égypte ancienne). De nos jours, environ 25% des grands barrages ont pour but premier la production d'énergie ou en tout cas comme but principal. Comme la charge hydraulique est un des paramètres importants pour générer de l'énergie, il est logique de constater qu'une grande part des grands barrages a pour raison d'être l'hydroélectricité. Ainsi, l'hydroélectricité est le but premier pour plus de 80% des barrages de plus de 200 m de haut. Les dix plus grands barrages au monde ont tous en premier lieu, une vocation hydroélectrique comme le montre le Tableau 1.1.

1.   INTRODUCTION

Dams store water and this may address many purposes and functions. Storage may serve to raise the water level in a river over a particular section to facilitate navigation or to create head for hydropower generation. Storage may also be used to collect surplus flow during floods for later use in supplies for domestic, industrial, agricultural, ecological or hydropower generation needs. In certain circumstances it may be necessary to keep the storage level low to facilitate flood mitigation downstream of the dam. In addition to these many, often conflicting requirements for creating storage, reservoirs may offer significant recreational or touristic potential. Dams need to be designed and operated to ensure that they meet all the anticipated purposes in an economically and environmentally sustainable manner.

Dams are generally built on rivers. As a result, the dam structures may cause potentially serious discontinuities in natural river systems, particularly with regard to sediment transportation, fish migration and breeding patterns. The changed flow regime downstream of dams, as well as the new shoreline created inside a reservoir also impacts the riverine environment. Thus, environmental considerations often create operating restraints or requirements.

Dams, being structures built in rivers, also have to be able to safely withstand the full onslaught of the river over very long periods of time, including the impacts of large floods and sediment loads.

More often than not dams are designed, built and operated to serve more than one of the above functions. The various roles may have conflicting design and/or operational requirements, so a cost-effective workable compromise is often necessary. This situation is often further complicated when multiple dams are built downstream of each other on one river (a cascade). In many countries the complexity is further increased by inter-basin water transfer that then requires integrated operation of reservoirs in multiple river basins. Various complex approaches to systems analysis have been developed over the years to reliably simulate, optimise and control such operations.

As a result of the wide range of water uses requiring storage in reservoirs, integrated planning and operation of water resources is well-developed in many countries across the world, especially the storage reservoirs.

1.1.   THE ROLE OF DAMS

Dams have been designed, constructed and operated as integral parts of water resources management and development worldwide. Historically, most dams were built with irrigation supply as the primary purpose (from Babylonian or Egyptian times). Nowadays about 25% of large dams have hydropower as the primary purpose, or as one of their main purposes. As head is the primary concern when generating hydropower, it follows logically that a large portion of high dams have been built for hydropower development, with hydropower been the primary purpose for more than 80% of all dams higher than 200m. The ten highest dams in the world all have hydropower as a very important purpose as shown in Table 1.1 below.

Table 1.1
Liste des dix plus hauts barrages au monde

Nom du barrage	Pays	Hauteur (m)	Type	But principal
Jinping-1	Chine	305	Voûte	Hydroélectricité
Nurek	Tadjikistan	300	Digue	Hydroélectricité
Xiaowan	Chine	292	Voûte	Hydroélectricité
Xiluodu	Chine	285.5	Voûte	Hydroélectricité
Grand Dixence	Suisse	285	Poids	Hydroélectricité
Enguri	Géorgie	272	Voûte	Hydroélectricité
Manuel Moreno Torres	Mexique	261	Digue	Hydroélectricité
Nuozhadu	Chine	261	Digue	Hydroélectricité
Tehri	Inde	260.5	Digue	Hydroélectricité
Mauvoisin	Suisse	250	Voûte	Hydroélectricité

Clairement, barrages et hydroélectricité sont fortement liés en raison de la grande efficacité des barrages à stocker de l'eau et donc de l'énergie ou de futures ressources en eau. L'importance d'une exploitation intégrée des réservoirs et aménagements hydroélectrique en ressort, bien que celle-ci doive tenir compte également des besoins en eau, en soutien d'étiage, en navigation et autres besoins comme le transport des sédiments.

1.2. LES SOURCES D'ÉNERGIE

L'hydroélectricité et les grands barrages assurent déjà une part significative de la fourniture en énergie dans plusieurs pays. La puissance hydroélectrique installée de plusieurs pays est indiquée dans le Tableau 1.2, en regard du pourcentage que l'hydroélectricité représente vis-à-vis de la puissance installée totale.

Table 1.2
Puissance hydroélectrique installée dans quelques pays en 2014

Pays	Puissance hydroélectrique installée (GW)	Part de l'hydroélectricité par rapport à la puissance installée totale (%)
Chine	301.8	22
États-Unis	101.6	8
Brésil	89.3	70
Canada	77.7	63
Japon	49.7	17
Russie	50.4	21
France	25.4	10
Suisse	15.6	60

Source: Statistiques IHA

Table 1.1
List of the highest dams in the world

Dam Name	Country	Height (m)	Type	Primary Purpose
Jinping-1	China	305	Concrete arch	Hydropower
Nurek	Tajikistan	300	Embankment	Hydropower
Xiaowan	China	292	Concrete arch	Hydropower
Xiluodu	China	285.5	Concrete arch	Hydropower
Grand Dixence	Switzerland	285	Concrete gravity	Hydropower
Inguri	Georgia	272	Concrete arch	Hydropower
Manuel Moreno Torres	Mexico	261	Embankment	Hydropower
Nuozhadu	China	261	Embankment	Hydropower
Tehri	India	260.5	Embankment	Hydropower
Mauvoisin	Switzerland	250	Concrete arch	Hydropower

Clearly dams and hydropower are closely related because dams are very effective in storing water and thereby energy for future water resources. It follows that the integrated operation of hydropower stations and reservoirs is a primary concern, albeit as part of a greater integrated role for dams in water supply, navigation and the like.

1.2.   THE ENERGY POOL

Hydropower and associated large dams currently make up significant portions of the national power supply of various countries. The installed hydropower capacity of countries with abundant waterpower resources are shown in the table below:

Table 1.2
Installed hydropower capacity of selected countries (by the end of 2014)

Country	Installed hydropower capacity (GW)	Hydropower as percentage of installed capacity (%)
China	301.8	22
United States	101.6	8
Brazil	89.3	70
Canada	77.74	63
Japan	49.67	17
Russian Federation	50.47	21
France	25.37	10
Switzerland	15.61	60

Notes: From IHA Statistics

Traditionnellement la construction d'aménagements de pompage-turbinage a montré un certain retard par rapport aux aménagements hydroélectriques conventionnels. Ce retard relatif résulte du fait que les installations de pompage-turbinage sont en fait des consommateurs net, opposés aux producteurs nets. Mais le concept qui sous-tend ces aménagements est d'utiliser de l'énergie à bas prix en période de faible demande électrique, pour accumuler de l'eau, puis pour produire de l'énergie avec cette eau en période de forte demande électrique. C'est donc un concept purement économique, qui a conduit au développement de telles installations. De plus, les installations de pompage-turbinage présentent des caractéristiques telles qu'elles deviennent un élément essentiel de stabilisation du réseau électrique. Par exemple, au début des années 1930, la première installation de pompage-turbinage était construite au Lac Noir dans l'est de la France afin d'équilibrer consommation et production d'énergie. On reviendra sur ce concept plus tard. Récemment, plusieurs grands aménagements de pompage-turbinage ont été construits de par le monde comme l'illustre le Tableau 1.3.

Table 1.3
Liste de grands aménagements de pompage-turbinage dans le monde

Pays	Aménagement	Puissance installée (MW)
États-Unis	Bath County	3003
Chine	Huizhou	2448
Chine	Guangzhou	2400
Japon	Okutataragi	1932
États-Unis	Ludington	1872
Chine	Tianhuangping	1836
France	Grand Maison	1800
Afrique du Sud	Ingula	1300
Russie	Zagorskaya-1	1200
Suisse	Linth-Limmern	1000
Suisse	Nant de Drance	900

L'aversion vis-à-vis des grands barrages, qui a prévalue dans les années 1980 et les critiques face aux Crédits Carbone ont conduit à considérer que les aménagements au fil de l'eau produisaient de l'énergie renouvelable, durable et sans dommages pour l'environnement alors que l'hydroélectricité produite à partir de grands barrages était non-renouvelable, non-durable et impactant l'environnement. Cette perception, bien que déclinante perdure. Les concepts de développement durable sont en plein essor, atténuant les impacts environnementaux et sociaux tout en optimisant la production d'énergie.

L'émergence d'énergie issue de technologies "vertes" économiquement abordable sous forme photovoltaïque ou éoliennes, ou encore, dans une moindre mesure, d'aménagements au fil de l'eau, a fait augmenter la variabilité stochastique de la production dans beaucoup de réseaux électriques nationaux ou régionaux. Une telle variabilité dans un réseau peut être gérée seulement tant que le réseau présente une certaine flexibilité dans sa structure, pour compenser les fluctuations introduites par les "énergies vertes". Des réponses rapides ne sont possibles que par l'hydroélectricité, les turbines à gaz, et à moindre échelle, par des batteries.

Dans les pays qui présentent une grande dépendance de leur production de base aux centrales thermiques ou lorsque le potentiel de production hydroélectrique est limité, il deviendra de plus en plus important d'avoir recours à des aménagements de pompage-turbinage hydraulique ou a des turbines à gaz, pour compenser la production non-prévisible introduite par le développement des "énergies vertes".

Traditionally the construction of pumped storage hydropower stations has lagged behind development of conventional hydroelectric plants (HPP). This relative lack of development results from the fact that pumped storage schemes are net users of energy as opposed to net generators but the idea was to use cheap energy and store water at period of time where there is no electric need, and to produce energy with this water at period of strong electric consumption, it was an economic concept that lead to pumped storage power plants. Moreover, they have essential characteristics that make them a very useful tool for stabilizing the electric grid. For example, in the beginning of 1930, the first pumped storage HPP was built at 'Lac Noir' in the eastern part of France, to regulate energy generation and consumption. This aspect will be returned to later. More recently, many large pumped storage schemes have been built. Table 1.3 shows a sample of large pumped storage hydropower schemes around the world.

Table 1.3
List of large pumped storage schemes around the world

Country	Schemes	Installed capacity (MW)
United States	Bath County	3003
China	Huizhou	2448
China	Guangzhou	2400
Japan	Okutataragi	1932
United States	Ludington	1872
China	Tianhuangping	1836
France	Grand Maison	1800
South Africa	Ingula	1300
Russian Federation	Zagorskaya-1	1200
Switzerland	Linth-Limmern	1000
Switzerland	Nant de Drance	900

The negative sentiment surrounding large dams in the 1980's and the severe scrutiny and criticism of the Carbon Trading Program brought about the concept that runoff hydro development was renewable, sustainable and environmentally friendly, whereas hydropower schemes involving large dams were non-renewable and environmentally damaging. This perception still lingers today, although the sentiment is declining. Concepts of sustainable development are continuously developing, mitigating environmental and social impacts whilst optimising energy production.

The emergence of more affordable green energy technology in the form of photovoltaic, wind turbines and, to a lesser extent, runoff hydropower schemes has seen the introduction of increased stochastic variability in many national and regional power networks. Introduction of such variability into a grid can be managed only as long as the grid has sufficient flexibility in its current makeup to compensate for fluctuations introduced by green technologies. Such a rapid response is to be found only in hydropower, gas turbines and on a smaller scale, batteries.

In countries where there is large scale reliance on base load generation using thermal and/or nuclear power plants, and/or where there is limited hydropower potential, it will become increasingly important to consider the development of pumped storage hydropower or gas turbines stations to compensate for the unpredictability introduced by green energy development.

2. LES CRITÈRES D'EXPLOITATION DES RÉSERVOIRS

Les critères d'exploitation des réservoirs doivent tenir compte de quatre enjeux distincts : les conditions hydrométéorologiques, la sécurité des barrages et les mesures d'urgence, la gestion des crues et les besoins des usagers de l'eau. Chacun de ces thèmes est abordé ci-après.

2.1. CONDITIONS HYDROMÉTÉOROLOGIQUES

La prévision hydrométéorologique est le fondement principal sur lequel repose la gestion des réservoirs. Les ressources hydriques sont tributaires de la météorologie ; bien qu'une relation existe entre eux, les précipitations et le ruissellement ont tous deux un caractère stochastique. On peut obtenir une prévision pour un certain intervalle de confiance (par des méthodes stochastiques) jusqu'à 14 jours au moyen de modèles hydrologiques (qui transposent les précipitations en ruissellement). Au-delà de 14 jours, il faut s'en remettre à un domaine de recherche marqué par une approche statistique. Un modèle météorologique général, produit par quelques centres de recherche dans le monde, étend l'horizon de prévision jusqu'à deux mois, mais au prix d'une incertitude nettement accrue. Des études et recherches sont nécessaires pour étendre la prévision météorologique à plus long terme.

2.1.1. Rôle du ruissellement

La modélisation des réservoirs ne peut pas produire de résultats fiables et aucune optimisation d'exploitation n'est possible si l'on ne dispose pas de séries chronologiques fiables des apports d'eau.

Dans certaines parties du monde, les données historiques de ruissellement couvrent une longue période. Le plus souvent, toutefois, la période couverte est trop courte, ou les données comportent trop de valeurs manquantes ou douteuses pour être utiles ; il s'agit alors de savoir si l'on peut corriger de telles données ou les extrapoler sur une période plus longue. Souvent, l'historique de précipitations est plus long et plus fiable que celui de ruissellement. Plusieurs modèles de prévisions hydrologiques ont été élaborés et calibrés dans différents pays. Ces modèles peuvent se révéler des outils extrêmement utiles pour augmenter l'étendue temporelle d'un historique de ruissellement. Il faut toutefois les utiliser avec prudence hors des régions pour lesquelles ils ont été calibrés. Par exemple, un modèle élaboré et étalonné pour prédire les apports dans une région semi-aride, donnera vraisemblablement des résultats tout à fait erronés dans un bassin versant dont une part substantielle du ruissellement est liée à la fonte des neiges. Ici le contraste est évident, mais des différences plus subtiles peuvent influer notablement sur les résultats du modèle.

Si l'on dispose d'un historique de ruissellement suffisamment long, ou si des données couvrant une période limitée ont pu être extrapolées sur une période suffisamment longue (idéalement 70 ans ou plus), divers générateurs stochastiques peuvent être utilisés pour produire de multiples séries synthétiques d'apports, pouvant alimenter les modèles de gestion des réservoirs. Ces générateurs produisent des séries synthétiques qui préservent certaines caractéristiques de la série originale : en général le ruissellement annuel moyen, l'écart-type des apports et d'autres caractéristiques. Les multiples séries d'apports ainsi produites permettent d'attribuer un degré de confiance aux résultats des modèles de gestion des réservoirs. Là encore, il faut se garder d'appliquer un modèle que l'on ne comprend pas, ou d'utiliser un modèle hors de sa région d'origine sans d'abord s'assurer qu'il demeure valide.

2. RESERVOIR OPERATING CRITERIA

Generally, reservoir operating criteria are determined based on hydro-meteorological considerations, dam safety and emergency responses, flood management operations and water user requirements. Each of them is elaborated on in the following.

2.1. HYDRO-METEOROLOGICAL CONSIDERATIONS

Hydro-meteorological forecasting is the foundation for reservoir operation. Water resources depend on meteorological conditions; hence both rainfall and runoff are stochastic. Prediction may be made up to 14 days by means of hydrological models (which transform precipitation into runoff). On the long term after 14 days, this is still a research topic and the domain of statistical approach. General meteorological model make prediction up to two months but of course with a great amount of uncertainty. Sufficient research and studied are still needed to extend the prediction period.

2.1.1. The role of runoff

The optimal operation of reservoir is depended on the reliable inflow time series. Otherwise, reservoir modelling will never produce reliable results.

In certain parts of the world, long runoff records are available. However, more often than not, runoff records are either too short or include too many missed or unreliable readings. The question then becomes how such records can be corrected and/or extended. Usually, rainfall records are longer and more reliable available than runoff records. So various hydrological models which transform rainfall into runoff are developed and applied in different countries. They can be extremely powerful tools to extend limited runoff records, but must be used with caution outside the areas where they were developed and proved. For example, a model that has been developed and proven in a semi-arid region may provide completely wrong results in a catchment that generates a substantial part of its runoff from snowmelt. The contrast seems obvious and more subtle differences may affect model results significantly.

If a sufficiently long observed/recorded runoff record is available, or a short record has been extended into a sufficiently long synthetic record (typically about 70 years or more), there are various stochastic generators that can be used to generate a multiple synthetic time series of flow that can be used as input into system simulation models. The generated synthetic sequences keep certain statistical characteristics of the source series. Typically, the mean annual runoff, standard deviation of inflows and the like will be maintained. Multiple time series of flow generated by such means allow assurances to be assigned to outcomes from simulation models. Again, the caution holds not to use a model that the user does not understand and not to use models outside the area in which they were developed without verifying that the model remains valid.

2.1.2. *Prévision hydrologique et météorologique*

Le processus de prévision hydrologique et météorologique consiste à établir une évaluation qualitative ou quantitative des conditions hydrologiques et météorologiques futures pour un certain plan d'eau, secteur ou station hydrologique, d'après des informations hydrologiques et météorologiques observées ou actuelles. Des prévisions hydrologiques précises et disponibles au moment opportun, sont le fondement scientifique de la prise de décision pour la gestion des crues, la gestion optimale des ressources hydriques et l'exploitation des centrales hydroélectriques. Ces prévisions jouent un rôle crucial pour assurer la sécurité des personnes et des biens, pour favoriser la performance des projets hydroélectriques et pour soutenir la stabilité sociale et le développement durable.

Depuis les années 1930, la technologie de la prévision hydrologique a fait des pas de géant, avec au départ des formules empiriques et des modèles globaux, pour aboutir à des modèles distribués qui produisent des résultats remarquables. Grâce aux progrès des technologies de l'information et informatiques, la prévision hydrologique continue d'évoluer de façon très dynamique.

Parallèlement à cette évolution, la continuité des séries de ruissellement des rivières a été affectée par une croissance socioéconomique soutenue, par l'exploitation accrue des ressources hydriques et par l'aménagement d'un nombre croissant de réservoirs et de centrales hydroélectriques dans les bassins versants. En particulier, des réservoirs dotés d'une capacité de régulation saisonnière ont largement altéré les caractéristiques naturelles de ruissellement des systèmes hydrologiques. En outre, les changements climatiques modifient la répartition spatiale et temporelle des précipitations ainsi que leur intensité ; les caractéristiques des éléments hydrologiques comme l'évaporation, le ruissellement et l'humidité des sols ont changé ; les sécheresses et les inondations sont devenues plus fréquentes. Ainsi, les changements climatiques à l'échelle mondiale, l'aménagement de réservoirs en cascade et une croissance socioéconomique rapide entraînent une pression considérable sur le développement et l'utilisation des ressources hydriques dans les bassins versants. Ils suscitent une série de nouvelles propositions dans différentes disciplines et présentent de nouveaux problèmes et défis pour la prévision hydrologique.

2.1.2.1. Prévision météorologique

Actuellement, la prévision météorologique porte sur les situations météorologiques et sur les éléments météorologiques. Dans le premier cas, il s'agit de prévoir les déplacements, les changements d'intensité, la formation et la disparition de systèmes météorologiques (anticyclones, dépressions, crêtes et creux, surfaces frontales, etc.) ; dans le deuxième cas, la prévision porte sur des éléments météorologiques comme la température de l'air, la pression atmosphérique, l'humidité, la visibilité, les vents, la nébulosité, les précipitations, ainsi que les phénomènes météorologiques. Les deux sont en fait étroitement associés. Les situations météorologiques constituent la base de la prévision des éléments météorologiques. Si l'on considère la période de prévision météorologique, on distingue trois horizons : le court terme, le moyen terme et le long terme. En règle générale, la prévision comprise entre les prochaines heures et 1 à 3 jours est dite à court terme, la prévision comprise entre 3 et 14 jours est dite à moyen terme, et au-delà de 14 jours (prévisions mensuelles, trimestrielles et annuelles) on parle de prévision à long terme.

Actuellement, les organismes de prévisions météorologiques recourent à trois méthodes principales de prévision : la carte météorologique, la prévision numérique et la méthode statistique mathématique. Les deux premières méthodes sont surtout utilisées pour la prévision à court terme, et la troisième principalement pour la prévision à long terme. Dans la pratique, on combine ces trois méthodes, qui se renforcent mutuellement. Il importe aussi de tenir compte de la capacité des prévisionnistes de comprendre les situations météorologiques (la prévision ne doit pas être tirée d'une « boîte noire ») et aussi, comme il a été dit plus haut, de combiner la capacité des modèles et du personnel d'anticiper ou de prédire les situations.

2.1.2. Hydrological and meteorological forecasting

Hydrological and meteorological forecasting is the qualitative or quantitative prediction of weather and hydrological conditions of a certain water body, area or hydrological station over a certain period of time in the future on the basis of prior or current hydrological and meteorological data and information. Accurate and timely hydrological forecasting is the scientific basis of decision-making for flood control, optimal operations of water resources, and management operation of hydroelectric projects. It plays a crucial role in ensuring the safety of lives and property, enabling hydroelectric projects to deliver intended results, and facilitating social stability and sustainable development.

Since the 1930s, hydrological forecasting technology has advanced rapidly from empirical formulas and lumped models to distributed models, with remarkable results achieved. Thanks to the development of computer, communication and information processing technologies, hydrological forecasting is still developing strongly.

However, the continuity of rivers has been changed with socioeconomic growth, increased exploitation of water resources and a rising number of reservoirs and hydropower stations that are under construction in watersheds. In particular, reservoirs with seasonal regulating capacity have, to a large extent, altered the characteristics of the natural runoff of water systems in watersheds. Moreover, climate change is altering the spatial and temporal distribution and intensity of precipitation; the characteristics of hydrological elements such as evaporation, runoff and soil moisture have changed; and droughts and floods have become more frequent. Thus, global climate change, construction of cascade reservoirs and rapid socioeconomic growth have put tremendous pressure on the development and utilization of water resources in watersheds, and both spawned a series of new propositions in various disciplines and posed new problems and challenges to hydrological forecasting.

2.1.2.1. Weather forecasting

Currently, weather forecasting includes the forecasting of weather situations and the forecasting of meteorological elements. The former involves the forecasting of the movements, intensity changes, formation and disappearance of weather systems (high pressure, low pressure, troughs and ridges, frontal surface, etc.), while the latter involves the forecasting of meteorological elements such as air temperature, air pressure, humidity, visibility, wind, clouds and precipitation, as well as weather phenomenon. The two are closely related. Weather situations are the basis of the forecasting of changes of meteorological elements. In terms of duration of the forecast period, weather forecasting is divided into short-term forecasting, medium-term forecasting, and long-term forecasting. As a general rule, forecasting for the next few hours to 1 to 3 days is classified as short-term forecasting; forecasting for 3 to 14 days is classified as medium-term forecasting; and monthly, quarterly and yearly forecasting covering more than 14 days is classified as long-term forecasting.

Currently, weather stations employ three major weather forecasting methods: the weather chart method, the numerical forecasting method and the mathematical statistics method. The first two methods are mainly used in short-term forecasting, and the last one is primarily devoted to long-term forecasting. In actual forecasting, the three methods are combined to supplement one another.

Appliquée à l'exploitation des centrales hydroélectriques en cascade, la prévision météorologique sert surtout à produire les données de base sur les précipitations en vue de la prévision hydrologique, qui détermine les apports d'eau attendus aux réservoirs et les débits entre ceux-ci. Certaines conditions météorologiques exceptionnelles (précipitations abondantes, typhons, orages violents, blizzards, tempêtes de neige, pluies sur neige fondante, etc.) peuvent représenter une menace potentielle pour les barrages ou les lignes électriques, d'où la nécessité de les prévoir.

2.1.2.2. Prévision hydrologique

Selon son objet, la prévision hydrologique se divise en :

- Prévision du niveau de l'eau ;

- Prévision du ruissellement, y compris celle des crues et des étiages ;

- Prévision de l'état des glaces et de la couverture neigeuse ;

- Prévision du transport des sédiments ;

- Prévision de la qualité de l'eau.

Selon la durée de la période de prévision, la prévision hydrologique se divise en :

- Prévision à court terme (moins de 3 jours) ;

- Prévision à moyen terme (entre 3 et 14 jours) ;

- Prévision à long terme (15 jours ou plus).

Plusieurs méthodes sont utilisées à différentes étapes de la prévision hydrologique selon les différentes conditions aux limites. Dans les calculs de ruissellement, les méthodes d'usage courant comprennent celles de l'écoulement sur surfaces saturées, de l'écoulement par dépassement de la capacité d'infiltration et de l'écoulement mixte. Dans les calculs de confluence, les méthodes courantes sont notamment celles des lignes d'unité et des lignes de temps d'écoulement uniforme. Dans les calculs de cours d'eau, les méthodes courantes comprennent la méthode de la longueur de rivière caractéristique, la méthode Muskingum, la méthode de niveau (débit) de rivière et l'algorithme hydraulique de crue de rivière. Les modèles hydrologiques apportent une aide technique importante pour la prévision hydrologique. En général, selon la robustesse et la complexité des schémas physiques des écoulements d'eau, les modèles hydrologiques sont divisés en modèles empiriques, en modèles conceptuels et en modèles à base physique. Selon leur capacité à représenter la variation spatiale des écoulements d'eau, les modèles hydrologiques sont divisés en modèles globaux, en modèles distribués, ou en modèles semi-distribués. Actuellement, les principaux modèles hydrologiques conceptuels comprennent les modèles Xin'anjiang, TANK, Sacramento et HBV ; parmi les principaux modèles hydrologiques distribués et semi-distribués, citons les modèles SWAT, TOPMODEL, SHE et VIC. Ces modèles sont utilisés largement à travers le monde pour réaliser les prévisions météorologiques.

Pourtant différents pays ont adopté des modes de classification différents. Par exemple au Brésil, ces modèles sont classifiés selon deux types, les modèles stochastiques et les modèles physiques pluie-débit. Ces modèles sont adoptés par presque tous les exploitants hydroélectriques.

Weather forecasting applied to the operation of cascade hydropower stations mainly serves the purpose of precipitation forecasting, which provides basic rainfall information and data for hydrological forecasting in order to predict inflows to reservoirs and the flows in between. Some extreme weather may have serious effects on a dam, so it is also necessary to forecast meteorological events like rainstorm, typhoon, thunderstorm, blizzard and snow etc.

2.1.2.2. Hydrological forecasting

By content, hydrological forecasting is divided into:

- Forecasting of water level.

- Forecasting of runoff, including forecasting of floods and forecasting of low flows.

- Forecasting of ice conditions.

- Forecasting of sediment situations.

- Forecasting of water quality.

By duration of the forecast period, hydrological forecasting is divided in:

- Short-term forecasting, covering less than 3 days.

- Medium-term forecasting, covering 3 to 14 days.

- Long-term forecasting, covering 15 or more days.

Different methods are employed at different stages of hydrological forecasting on the basis of different boundary conditions. In the runoff yield process, frequently used methods include the method of runoff yield under saturated storage, the method of runoff yield under excess infiltration, and the method of mixed runoff yield. In the confluence process, frequently used methods include the method of unit lines and the method of uniform flow timelines. In the watercourse computation process, frequently used methods include the characteristic river length method, the Muskingum method, and the corresponding river level (flow) method, and the hydraulic river flooding algorithm method. Hydrological models provide strong technical support for hydrological forecasting. In general, these existing hydrological models are divided into empirical models, conceptual models, and physical models, according to the robustness and complexity of the physical patterns of the movements of water flows. Hydrological models are divided into lumped hydrological models and distributed hydrological models according to their ability to reflect the spatial changes of the movements of water flows. Between the two types of modes, there are the semi-distributed models. At present, the main conceptual hydrological models include the Xin'anjiang model, the TANK model, the Sacramento model, and the HBV model while the main semi-distributed and distributed hydrological models include the SWAT model, the TOPMODEL, the SHE model, the VIC model. These models are used to the greatest extent to make the forecasting across the world.

However, different countries have various ways of models' classification. For example, in Brazil, these models are generally classified into two types, called stochastic models and rainfall-runoff models, and are adopted by almost all the hydropower stations.

Puisque la production d'électricité est tributaire des conditions météorologiques et hydrologiques, il doit exister un service ou du personnel responsable de ces prévisions. Prenons par exemple le fleuve Kiso au Japon : une division particulière est chargée des prévisions de débit à partir des prévisions météorologiques publiées par l'agence nationale de météorologie. Autre exemple, le fleuve Yang Tsé en Chine pour lequel les prévisions météorologiques et de production hydroélectrique sont confiées à une division spéciale. Il est également important que les prévisionnistes soient qualifiés au sens de la compréhension des phénomènes météorologiques, des phénomènes hydrologiques et de l'utilisation des modèles de simulation et de prédiction.

De nos jours, l'exactitude de la prévision hydrométéorologique – surtout à moyen et à long termes – ne suffit pas aux besoins de régulation conjointe des réservoirs et des centrales hydroélectriques, de sorte que des améliorations sont nécessaires. Des percées futures pourraient provenir des technologies de prévision numérique, de couplage hydrométéorologique ou de prévision hydrologique à moyen et à long terme.

2.2. SÉCURITÉ DES BARRAGES ET MESURES D'URGENCE

L'exploitation en sûreté des barrages et des réservoirs connexes est un impératif essentiel. L'évaluation de la pertinence et de la précision des prévisions météorologiques, comme exposé précédemment, peut avoir un impact sur la sûreté du barrage ou du réservoir. Dans le cas peu probable d'une situation d'urgence fortuite, il est essentiel d'avoir en place des systèmes d'intervention efficaces et efficients. Ces deux aspects sont exposés brièvement ci-après.

2.2.1. Sécurité des barrages

Les barrages sont conçus pour être stables et demeurer sûrs dans toutes les conditions possibles : charges mécaniques, contraintes thermiques, événements sismiques, etc. En outre, ils doivent pouvoir absorber les crues naturelles avec leurs charges de sédiments et de corps flottants. Ces conditions font toutes l'objet de divers bulletins CIGB, par exemple :

- L'érosion interne dans les barrages et digues en exploitation et dans leur fondation (Bulletin 164, 2015)

- Gestion intégrée du risque de crue (Bulletin 156, 2014)

- Gestion de la sécurité des barrages en exploitation (Bulletin 154, 2017)

- Rôle des barrages dans l'atténuation des crues – Synthèse (Bulletin 131, 2006)

Les exigences relatives à la sécurité des barrages sont également encadrées par la loi dans divers pays.

L'exploitation sécuritaire d'un réservoir nécessite de prendre en compte la stabilité de la structure du barrage, la stabilité des berges en cas de variation rapide du niveau de l'eau et l'impact sur l'environnement en aval.

2.2.2. Surveillance des barrages

Les activités de surveillance sont essentielles pour vérifier que le barrage et son réservoir respectent les exigences de conception. La surveillance des barrages est encadrée par la loi dans de nombreux pays, et fait également l'objet de recommandations dans les bulletins CIGB pertinents.

Since hydropower production depends on weather and runoff, there should be special departments or specialists responsible for it. Take Japan's Kiso River as an example, there is a specific division making runoff forecasts based on weather forecasting data published by the meteorological agency. Furthermore, in China's Yangtze River, both meteorological and hydrological forecasts are made by one special department. It is also important that forecasters should be qualified with understanding meteorological and hydrological situations and utilizing modals to simulate and predict.

Nowadays the accuracy of hydro-meteorological forecasting cannot meet the demands of integrated operation of hydropower stations and reservoirs, especially for medium-term and long-term forecasts, so it needs further development. Future breakthroughs can focus on numerical value forecast technology, hydro-meteorological coupling prediction technology, medium-term and long-term hydrologic forecast technology.

2.2. DAM SAFETY AND EMERGENCY RESPONSE

Operating dams and the associated reservoirs in such a way as to ensure dam safety is the primary concern. Evaluating the pertinence and accuracy of hydrological predictions as mentioned in previous sections is important to ensure a feedback on these predictions, to be able to improve them, since they may have an impact on the safety of the dam and reservoir. In the unlikely event of an unforeseen emergency occurring, it is essential to have practical, effective and efficient response systems in place. These two aspects are briefly outlined in the following.

2.2.1. Dam safety

Dams are designed to be stable and remain safe when subjected to all conceivable loads and conditions including physical loads, thermal loading and seismic events. In addition, dams need to be able to handle natural floods and the associated sediment and/or debris loads. These are all subjects to various ICOLD bulletins such as:

- Internal Erosion of Existing Dams, Levees and Dikes, and their Foundations (Bulletin 164, 2015)

- Integrated Flood Risk Management (Bulletin 156, 2014)

- Dam Safety Management: Operational Phase of the Dam Life Cycle (Bulletin 154, 2017)

- Role of Dams in Flood Mitigation - A Review (Bulletin 131, 2006)

Dam safety requirements are also regulated by laws in various countries.

Safe operation of a reservoir includes consideration of the stability of the dam structure, reservoir basin slope stability as affected by rapid water level variations and impact on the downstream environment.

2.2.2. Dam surveillance

Dam surveillance is fundamental to check and control that the dam and the associated reservoir behave as envisaged during their design. Dam safety is legislated in many countries. It is also highly emphasized in the relevant ICOLD bulletins.

En principe, l'auscultation et l'analyse de la sécurité d'un barrage sont de rigueur :

- À la mise en eau initiale ;

- Dans le cadre d'enquêtes de sécurité périodiques ;

- Après une crue exceptionnelle, un séisme important ou toute circonstance inhabituelle.

Lorsque requis, des mesures sont effectuées et les résultats transmis aux autorités concernées. L'auscultation d'un barrage porte notamment sur la déformation structurale, les infiltrations d'eau, la relation contrainte/déformation, la tenue hydraulique, les fortes activités sismiques, etc.

2.2.3. Mesures d'urgence

Les procédures d'urgence sont généralement définies à l'étape de conception du barrage. Les deux thèmes prédominants sont l'analyse de la rupture du barrage en cas de défaillance et la situation où le niveau d'eau du réservoir devrait être abaissée sur une période raisonnable en cas d'urgence structurale. De nombreux pays ont une législation à cet égard. Il est pertinent de noter qu'un abaissement d'urgence du niveau d'eau a des impacts sur les usagers de l'eau du réservoir ainsi que sur les usagers en aval, mais la sécurité demeure le critère primordial.

La prévision en temps réel lors situation d'urgences peut être difficile (inondation), voire impossible (secousse sismique). Ainsi lorsqu'une situation d'urgence se produit, elle doit être gérée par un personnel compétent et bien formé, qui doit alors être en mesure d'intervenir rapidement. Les mesures prises suivent normalement les plans d'urgence préétablis. Certains grands fleuves sont partagés par plusieurs pays et plusieurs organisations nationales sont en cause ; habituellement, des ententes internationales établissent la responsabilité opérationnelle.

En France, par exemple, deux grands principes sont mis en avant. Le premier consiste à ne pas hésiter à surréagir ; le second, à former une cellule de gestion de crise. Cette cellule devra coordonner :

- Les actions et interventions techniques ;

- Les équipes ou missions de sauvetage éventuelles ;

- Les communications au sein de la cellule ainsi qu'avec les entités extérieures touchées.

La cellule de gestion de crise doit disposer de tout le soutien technique et logistique nécessaire. Elle doit en outre être formée à dialoguer avec les médias. Les principes sont simples, mais leur application en temps réel est très difficile ; les équipes d'exploitation doivent recevoir des formations régulières.

2.3. BESOINS DES USAGERS DE L'EAU

L'exploitation des réservoirs est un élément clé pour la plupart des systèmes d'approvisionnement en eau. Les réservoirs fournissent de l'eau pour diverses utilisations consommatrices, principalement l'irrigation, la consommation domestique, les usages industriels et les besoins écologiques (y compris la pisciculture), et aussi pour diverses utilisations non-consommatrices comme la navigation, la production hydroélectrique et les activités récréatives. Habituellement, les réservoirs ne sont pas conçus pour toutes ces utilisations ; ils sont exploités en fonction de leur vocation principale, d'autres utilisations étant parfois prises en compte. En général, l'irrigation et la production hydroélectrique sont les utilisations de l'eau les plus importantes et les plus répandues.

Usually, monitoring and analysis of dam safety should be undertaken when:

- First impounding.

- During regular safety investigations.

- During special investigations after a huge flood, serious earthquake or any unusual circumstances.

When necessary, measures should be taken, and results should be reported to related authorities. Dam safety monitoring includes the monitoring of structure deformation, seepage, stress-strain, hydraulics, strong seismic activities etc.

2.2.3. Emergency response

Emergency response procedures are generally developed during the dam design phase. The two most common considerations are dam break analyses in the event of a catastrophic failure, and the need to be able to draw the reservoir water level down over a reasonable time in the event of some structural emergency. Many countries have legislation in this regard. It is worthwhile to note that such emergency drawdown may affect water users in the reservoir as well as those further downstream, but safety is the first stake to be taken into account.

Real time prediction of emergencies is either difficult (e.g. flash floods) or impossible (e.g. seismic event). If an emergency occurs, action must be taken quickly by well-trained competent people. Actions are normally taken in accordance with previously prepared emergency plans. These situations are normally managed by a designated organisation staffed by suitably qualified and experienced people. Some major rivers are shared between several countries with more than one national organization involved. Under these circumstances international agreements usually assign operative responsibility.

In France, as an example, there are two broad principles. The first principle is to overreact, and the second principle is to set up a crisis management team. The team will coordinate:

- Technical action and /or interventions.

- Possible rescue teams or missions.

- Communication inside the team as well as with externally affected parties.

The team should have all the necessary technical and logistical support required. In addition, they should be trained in handling the media. These principles are simple but real time application is very difficult, operating team has to be regularly trained.

2.3. WATER USER REQUIREMENTS

Operation of reservoirs is a key part of almost all water supply systems. Reservoirs provide water for various consumptive uses, mainly including irrigation, domestic use, industrial use, ecological demands (including fish-farming), and various non-consumptive uses, including navigation, hydropower generation, and recreation. Usually reservoirs are not designed for all of these uses, they are operated according to their primary purpose and sometimes other uses also will be taken into account. In general, irrigation and hydropower generation are the primary and most common types of water use from dams.

La priorité d'approvisionnement et les garanties d'approvisionnement pour les divers types d'utilisateurs sont encadrées par la loi dans de nombreux pays. Ces aspects sont évoqués ci-après.

2.3.1. Irrigation

La plupart des grands barrages ont l'irrigation pour principal objectif, ou du moins l'un des objectifs principaux. Les besoins d'irrigation varient généralement selon la saison et peuvent nécessiter une garantie d'approvisionnement relativement faible. En conséquence, la quantité d'eau fournie peut varier d'un mois à l'autre, et aussi d'année en année.

L'approvisionnement peut être effectué de diverses façons :

- Évacuation de l'eau par le barrage pour un prélèvement en aval, par exemple par un seuil déversant ;

- Déversement de l'eau du réservoir dans un canal, par gravité ou par pompage ;

- Par une conduite sous pression qui traverse la paroi du barrage.

Le mode de prélèvement de l'eau d'irrigation à partir du barrage peut influer sur le volume prélevé ou sur le moment choisi.

La variabilité des volumes prélevés pour l'irrigation, selon la saison et aussi selon le degré de sécheresse durant l'année, ainsi que les niveaux de tolérance relativement faible qui peuvent être assumés par les irrigateurs, font de l'irrigation la variable la plus importante dans l'optimisation de l'exploitation d'un réservoir. En outre, les débits relâchés pour l'irrigation peut influer sur le débit de la rivière en aval des zones irriguées.

La problématique de l'irrigation est présente dans le monde entier. En France, l'eau se fait plus rare en été ; pendant cette période, l'agriculture réclame une part de l'eau stockée dans les réservoirs en cascade de l'Ariège. Pour citer un autre cas en Chine, le bassin du fleuve Jaune connaît une forte demande d'eau pour l'irrigation, étant donné les fluctuations saisonnières du régime hydrique ; plusieurs réservoirs en cascade sont commandés de façon coordonnée pour répondre à la demande d'irrigation, surtout dans les secteurs amont et intermédiaire du fleuve. Les objectifs sur la chaine Volga-Kama en Russie, permettent de préserver les régions est de la Volga des sécheresses et ainsi d'assurer la production de céréales.

2.3.2. Consommation domestique

La consommation domestique par habitant est appelée parfois « demande primaire ». Dans sa forme la plus basique, elle correspond à l'eau consommée par les humains. Comme un manque d'eau potable peut menacer la vie humaine, les garanties d'approvisionnement, qui y sont associées, sont généralement élevées.

Les besoins en eau domestique par habitant augmentent généralement avec le niveau de vie de la population. Si la consommation domestique est élevée, une certaine variabilité dans l'approvisionnement (restrictions) peut être acceptable en période de sécheresse.

Bien que la consommation en eau soit rarement le premier besoin à satisfaire, c'est un besoin stratégique et il est donc important de le prendre en compte correctement lorsque l'on définit les règles d'exploitation d'un projet de réservoirs en cascade.

The priority of supply and the assurance of supply to the various types of users is the subject of legislation in many countries. These aspects are elaborated on below.

2.3.1. *Irrigation*

Most large dams have irrigation supply as their primary purpose, or as one of their main purposes. Such supply generally needs to vary seasonally and may require a relatively low assurance of supply. Accordingly, supply may vary from month to month and/or from year to year.

Supply can be effected by:

- The releasing of water from the dam to be abstracted further downstream, possibly from a weir.

- The releasing of water into a canal. The canal may be a gravity feed or may require some pumping from the reservoir into the canal.

- A pressure pipeline through the dam wall.

The means selected to abstract irrigation water from the dam may affect the extent or timing of drawdown of the reservoir that would be desired.

The variable rate of abstraction of irrigation water, both in terms of seasonality and in terms of the level of drought that may be experienced in any one year, together with the relatively low levels of assurance that can be tolerated by irrigators, make irrigation use the single biggest variable in reservoir operation optimisation. Furthermore, return flows from irrigation may affect river flows downstream of the irrigated areas.

The issue of irrigation is a challenge around the world. In France, water shortages during summertime. In this period there are agricultural demands that require water stored in the Ariege cascade reservoirs. While in the Yellow River Basin, China, the demands on water supply for irrigation are great. Due to the variable seasonal water regime, cascade reservoirs are operated together to meet the demands of irrigation especially in the upstream and midstream. The objectives on the Volga-Kama cascade in Russia also contribute a lot to the salvation of eastern Volga areas from periodic crop failures and ensure grain production.

2.3.2. *Domestic usage*

Per capital domestic use, sometimes referred to as primary demand, is water used for human consumption. In its most basic form, it represents drinking water. As failure to supply drinking water may threaten human life, levels of assurance of supply required are generally high.

Per capita domestic water demand generally grows as the affluence of those supplied increases. At higher levels of domestic supply some variability in supply (restrictions) may be acceptable in times of drought.

Although domestic supply is rarely the single biggest water user it is a strategic supply and thus important to account for properly when designing integrated reservoir operating rules.

La vallée de l'Ariège et le fleuve Tennessee remplissent tous deux une fonction importante d'approvisionnement en eau dans leurs bassins respectifs. Électricité de France (EDF) doit maintenir une station d'eau potable pour les villes voisines de la vallée de l'Ariège et doit superviser le traitement de l'eau avant son injection dans le réseau de distribution. Aux États-Unis, l'eau est prélevée en plus de 700 points le long du fleuve Tennessee et de ses affluents pour l'approvisionnement d'environ 4 millions de personnes.

Le projet de transfert d'eau du sud au nord en Chine est un autre cas, qui illustre les échanges d'eau entre bassins versants. Une fois le projet achevé, 44,8 milliards de m³ d'eau seront transférées chaque année vers le nord afin de venir en aide à plus de 7 millions de personnes actuellement contraintes de boire de l'eau saumâtre ou à teneur élevée en fluor. La source d'eau d'un des trois tronçons est un réservoir dont la capacité a été accrue par le rehaussement du barrage.

2.3.3. Usages industriels

Les volumes d'eau nécessaires à l'industrie peuvent être substantiels, avec souvent un niveau élevé de garantie d'approvisionnement. Les besoins stratégiques peuvent inclurent l'eau nécessaire au refroidissement des centrales électriques thermiques. L'approvisionnement en eau des mines et d'autres gros consommateurs industriels, notamment dans l'agroalimentaire, offre souvent les meilleures perspectives de rentabilité pour une infrastructure d'approvisionnement en eau. C'est pourquoi l'eau industrielle est souvent un argument essentiel pour la justification économique de nouveaux barrages. L'hydroélectricité tombe dans cette catégorie ; mais comme il s'agit d'une utilisation non-consommatrice d'eau, nous y reviendrons plus loin.

Dans le réseau de la Tennessee Valley Authority (TVA) aux États-Unis, le système de réservoirs fournit aussi l'eau de refroidissement des centrales nucléaires et au charbon de la TVA. Ces centrales, qui produisent une grande partie de l'énergie de la zone de desserte électrique de la TVA, sont dépendantes de l'exploitation des réservoirs. Comme la disponibilité de ces centrales est essentielle à la fourniture d'une électricité fiable et abordable, leur approvisionnement en eau de refroidissement est un objectif d'exploitation important pour le système de réservoirs. Les centrales nucléaires et au charbon exigent de grandes quantités d'eau de refroidissement. Le retour de cette eau dans le système de réservoirs après utilisation est réglementé (par un système de permis) afin notamment de limiter la hausse de température de l'eau des réservoirs qui peut en résulter. Cette réglementation vise à maintenir la qualité de l'eau et à protéger la vie aquatique. Les débits minimaux du système dans le fleuve Tennessee sont en partie tributaires des besoins en eau de refroidissement de ces centrales.

D'autres réservoirs ont aussi des besoins en eau industrielle, comme le barrage de Dez en Iran et la cascade de Volga-Kama en Russie. La situation est identique en France, dont les centrales sont à 80 % thermiques et notamment nucléaires ; le besoin d'eau de refroidissement devient particulièrement sensible avec l'incidence croissante de périodes de sécheresse et avec les changements climatiques. Ces besoins doivent être pris en compte dans la gestion des grands fleuves, où les centrales nucléaires et thermiques sont implantées – en particulier le Rhône, la Garonne et la Loire, trois des principaux fleuves de France.

2.3.4. Besoins écologiques

Les besoins écologiques en eau correspondent au minimum d'eau que doit évacuer un réservoir pour maintenir un écosystème fluvial sain en aval du barrage, notamment les besoins en eau de la faune aquatique et de son milieu de vie. Essentiellement, il s'agit du volume d'eau qui permet de préserver la stabilité dynamique de la biocénose et des habitats dans les systèmes écologiques. Dès la construction d'un réservoir, il convient de prévoir les débits d'évacuation minimaux nécessaires à cette fin. En général on met en place un débit minimal pour satisfaire les besoins en eau écologique, lors de la construction du réservoir. Diverses méthodes fort complexes d'évaluation des besoins écologiques en eau ont été élaborées dans différentes parties du monde et le respect de ces besoins est encadré par la loi dans de nombreux pays.

The Ariège valley and Tennessee River both play an important role in the water supply of their basins. Electricité de France (EDF) has to maintain a water station to provide water for cities around the Ariège valley and has to survey the treatment of water before introducing it into the distribution network. In the USA, water is withdrawn at over 700 points along the Tennessee River and its tributaries to the benefit of approximately 4 million citizens.

China's South Water to North Project is another typical case of trans-basin water transfer. With the completion of the project, 44.8 billion tons of water every year will be transferred to the north to improve the water quality for more than 7 million people. The water source of one of the three transfer lines is a reservoir which is expanded by heightening the dam to increase storage capacity.

2.3.3. Industrial usage

Industrial water use volumes can be significant and often demands high levels of supply assurance. Strategic supplies may include water supplies to thermal power stations and the like. Water supply to mines and other large industrial users, including the food processing industry, often yield the highest return on the investment in water supply infrastructure. Accordingly, industrial water use is often key to building an economic case for developing new dams. Hydropower falls under this category, but it will be addressed later since it is a non-consumptive user of water.

In the Tennessee Valley Authority (TVA) System in the USA, operation of the reservoir system also provides cooling water for TVA's coal and nuclear power plants. TVA coal and nuclear plants provide a large portion of the energy needed for the TVA power service area and therefore increased depend on reservoir operations. Because their availability is essential to TVA's ability to provide reliable, affordable electricity, support of coal and nuclear plant operations by the reservoir system is an important operating objective. It is also the case in France for the cooling of nuclear plants built on rivers.

Coal and nuclear plants require large quantities of cooling water to operate. Return of the cooling water to the reservoir system is regulated (by permit) and includes limitations on the increase in reservoir water temperatures that can result from power plant discharge. These limitations are established to maintain water quality and protect aquatic life. System minimum flows in the Tennessee River are governed in part by the cooling water needs of these plants. Other reservoirs also have the demand of industrial water, such as Dez Dam in Iran and Volga-Kama cascade in Russia.

2.3.4. Ecological requirement

Ecological water demand is the minimum water that is required to be released from a reservoir in order to maintain a healthy riverine eco-system downstream of a dam, including the water requirements of organisms themselves and the water requirements of the environment in which organisms live. In essence, ecological water demand is the volume of water required to maintain the dynamic stability of biocenosis and habitat in ecological systems. Generally, there should be regulation of the minimum discharge required for ecological water demands when a reservoir is constructed. Various highly sophisticated methods of determining ecological water demands have been developed in different parts of the world and supply of these demands are legislated in many countries.

Malheureusement, les débits écologiques sont insuffisants dans plusieurs pays, ce qui entraîne une détérioration de nombreux écosystèmes fluviaux. Souvent, ce manque d'approvisionnement en eau va de pair avec un contrôle déficient des effluents des usagers de l'eau, ce qui aggrave l'état écologique des cours d'eau. À l'évidence, le respect des besoins écologiques doit faire partie intégrante de l'élaboration des règles d'exploitation intégrée pour tout barrage. Ceci dans un contexte où les pays prennent de plus en plus conscience de l'enjeu du développement écoresponsable des barrages et des réservoirs. Il peut devenir plus difficile d'assurer de tels débits si les réservoirs sont disposés en série ou s'ils sont exploités en parallèle, sur des affluents adjacents d'un même réseau hydrographique ou sur des rivières adjacentes et lorsqu'il y a un quelconque transfert d'eau entre bassins.

La pisciculture exige un niveau d'eau stable – donc une évacuation et une mise en eau lentes – afin de prévenir l'asphyxie et la mort des poissons, de permettre aux poissons de frayer et de se déplacer, sans compter certains débits d'évacuation spéciaux obligatoires pour permettre la fraie en aval.

En fonction de leur situation particulière, beaucoup de pays ont des besoins en eau pour satisfaire les besoins écologiques. Au Canada et aux États-Unis, la gestion écologique de l'eau dans le fleuve Columbia vise à favoriser la migration du saumon. Pour atteindre cet objectif, plusieurs mesures techniques ont été mises en place, comme l'augmentation du débit d'eau relâché. Dans le cas du complexe des Trois-Gorges, afin de favoriser la fraie des « quatre grandes espèces » de carpes chinoises, on réhausse artificiellement le niveau de l'eau dans les cours intermédiaire et inférieur du fleuve Yang Tsé afin de simuler la crue naturelle des mois de mai et juin. Au Japon un débit de base est nécessaire pour maintenir le bon fonctionnement de la rivière. Les besoins en eau écologique du barrage de Karun en Iran visent à contrôler la qualité de l'eau de la rivière et de prévenir l'intrusion d'eau salé du golfe persique. Des mesures identiques sont adoptées dans plusieurs pays, spécialement en Europe, où la législation sur l'écologie a été renforcée.

2.3.5. Navigation

Le transport par bateau demeure l'un des modes de transport les moins coûteux pour les personnes et les biens, surtout pour les marchandises lourdes. De nombreux fleuves et rivières sont d'ailleurs d'importantes voies de transport. Bon nombre de barrages, équipés d'ascenseurs à bateaux ou d'écluses, sont construits soit pour approfondir le cours d'eau afin que de plus grands navires puissent y circuler, soit pour permettre la navigation plus loin en amont qu'il ne serait autrement possible.

Bien que la navigation ne soit pas une utilisation consommatrice d'eau, elle nécessite que les barrages construits à cette fin soient maintenus pleins ou presque pleins en tout temps. Cela peut représenter une contrainte importante, qu'il faut incorporer aux règles d'exploitation intégrée des réservoirs.

La navigation nécessite une profondeur d'eau garantie et des apports spéciaux pour les voies navigables, notamment celles qui mènent aux installations portuaires, aux chantiers navals et aux écluses. De brusques fluctuations du niveau de l'eau en amont et en aval des barrages nuisent à la navigation ; elles altèrent la stabilité des coupes de dragage, détériorent les conditions d'amarrage en hiver et obligent à des travaux de maintenance périodiques dans le chenal de navigation. Il convient donc d'envisager la construction de barrages de régulation en aval de grandes centrales hydroélectriques, pour démoduler le débit des turbines et réduire les effets néfastes et ainsi améliorer les conditions de navigation.

Prenons l'exemple du fleuve Mississippi aux États-Unis. L'aménagement de ce cours d'eau a toujours fait une grande place à la navigation, en particulier pour son cours intermédiaire et inférieur. La construction de barrages et d'autres projets auxiliaires sur le fleuve Mississippi en fait une voie de transport idéale, avec des profondeurs d'eau et des débits stables.

Sadly though, there is a lack of such supply in several countries, which has led to the deterioration of many riverine ecosystems. Often such lack of provision goes hand in hand with poor control of effluent from water users which further damages river ecosystems. It follows that the supply of ecological demands ought to be an integral part of developing integrated operating rules for any dam. This is also coming with the raising for conscience of countries about sustainable developing of dams and reservoirs. Allocating such supply provisions can become more difficult if reservoirs are built in series or if they are operated in parallel, either in adjacent tributaries in the same river system or in adjacent rivers if there are some form of inter-basin water transfer present.

Fish-farming requires a stable water level, slow drawdown and impoundment to prevent fish from suffocating and dying, to allow for spawning, and to allow fish to migrate, as well as special obligatory releases to allow for spawning in downstream.

According to their unique situations, many countries have the demand of ecological water. In Canada and the USA, the purpose of ecological regulation in the Columbia River is to help salmon to migrate. In order to accomplish this goal, different engineering measures, such as increasing the amount of water released from reservoirs are taken. The ecological regulation of the TGP, aimed at promoting the spawning of the "four major Chinese carps", refers to the artificial raising of the water level in the middle and lower reaches of the Yangtze River by simulating the natural water rise process in May and June. In Japan, the basic discharge is also necessary for maintaining regular functions of rivers. The environmental demand of Greater Karun River in Iran is for river water quality control and prevention of Persian Gulf saltwater intrusion. Identical measures are taken in other countries, especially in Europe where legislation has been reinforced.

2.3.5. Navigation

Shipping remains one of the most cost-effective means of transportation people and goods, particularly heavy goods. Accordingly, many rivers constitute important transport corridors. Often dams, which incorporate ship lifts or locks, are built to either deepen the rivers so that larger ships can be accommodated, or to allow navigation further up the rivers than would otherwise be possible.

Although navigation is not a consumptive water use, it does require that the dams built to facilitate such navigation be kept full or close to full at all times. This may present a significant constraint on reservoir operation, which has to be built into the integrated operating rules.

Guaranteed depths are required for navigation, as well as special releases for waterways, ways leading into ports, and shipyards and locks. Sharp fluctuations of water levels upstream and downstream of dams are unfavourable for navigation as they affect the stability of dredge cuts, worsen conditions for winter moorage and make periodic maintenance work necessary in the canal. Therefore, re-regulation dam construction should be considered downstream of large-scale hydropower stations for regulating the outflow from turbines in order to reduce adverse effects caused by outflow changes and improve navigation conditions.

Taking the Mississippi River in the USA as an example, high importance has always been placed on navigation throughout the development of the river, especially midstream and downstream. The construction of dams and other auxiliary projects has turned the Mississippi River into a 'golden way' with stable depth and flow.

Un exemple en Europe, le Rhin traverse plusieurs pays: la Suisse, la France et l'Allemagne, et est le principal moyen de transport de marchandises vers la Suisse (depuis l'Allemagne ou la France). Il existe donc des contraintes importantes pour assurer sa sécurité par une bonne maintenance de ces ouvrages, qui ont bien entendu une incidence sur les centrales hydroélectriques de la cascade du Rhin.

Autre exemple, cette fois en Chine : le fleuve Yang Tsé, appelé « la voie navigable d'or », est la voie fluviale la plus fréquentée du monde pour le transport des marchandises. Après la mise en eau de la centrale électrique des Trois-Gorges, les conditions de navigation entre Chongqing et le barrage se sont améliorées notablement, à cause de l'élimination de tous les hauts fonds et rapides. La profondeur de dragage de ce tronçon a été améliorée et certains affluents ont pu être ouverts à la navigation. En outre, l'effet régulateur du barrage de Gezhouba (GZB) en aval amoindrit l'impact causé par le débit d'évacuation irrégulier de la centrale des Trois-Gorges.

2.3.6. Production hydroélectrique

Comme il a été souligné en introduction, l'hydroélectricité est la vocation principale d'une proportion importante des grands barrages dans le monde. À mesure que la demande mondiale d'énergie augmentera, le besoin de barrages de grande hauteur, voués principalement à la production hydroélectrique, progressera assurément.

Bien qu'il ne s'agisse pas d'une utilisation consommatrice de l'eau, la production hydroélectrique exerce une influence variée et complexe sur la manière dont les réservoirs peuvent ou doivent être exploités. Les centrales hydroélectriques sont impliquées dans la fiabilité et la sécurité des ouvrages et équipements hydrauliques, l'utilisation efficiente de la ressource eau (en vue d'une production accrue) et l'augmentation des revenus liés à la vente d'énergie. Dans le cas de centrales en cascade, le débit doit être géré de façon à maximiser la production totale, tout en tenant compte des besoins des autres usagers de l'eau.

Certains cas mentionnés dans le bulletin ne traitent pas en détail de la production hydroélectrique, mais presque tous les aménagements en cascade sont liés à cette exigence. De plus, quelques aménagements en cascade ont été construits essentiellement pour la production d'énergie, comme la cascade d'Itaipu sur la rivière Parana. Elle joue un rôle important dans l'approvisionnement en énergie et promeut significativement le développement économique au Brésil, au Paraguay et en Argentine.

2.3.7. Activités récréatives

Les usagers d'activités récréatives ont besoin d'une certaine stabilité du niveau de l'eau dans la zone de stockage ou en aval du barrage, ceci pendant une certaine période de l'année ou saison (mais pas toute l'année).

Le canyonisme, le kayakisme et la navigation de plaisance sont pratiqués surtout en été ; les producteurs d'électricité doivent tenir compte de ces activités pour des raisons évidentes de sécurité. Il peut en résulter des contraintes ponctuelles, selon les conditions hydrologiques. En cas de travaux de maintenance dans une centrale, il importe d'avertir les usagers que les conditions de la rivière seront altérées pendant la période visée. Il faut aussi tenir compte du débit nécessaire pour les activités nautiques ; parfois, un accord contractuel prévoit le maintien d'un certain débit pendant la saison sèche ou pour assurer une activité minimale pendant l'été (développement durable).

En France et en Norvège, des besoins en eau sont établis pour des activités récréatives en aval de barrages. C'est particulièrement le cas dans la vallée de l'Ariège, où se pratiquent de nombreuses activités récréatives. Le canyonisme, le kayakisme et la navigation de plaisance connaissent une intensité particulière en été : certaines portions des rives servent de plage en été et un téléski aquatique est installé sur le lac de Garabet pendant l'été. Toutes ces activités entraînent des contraintes pour la production d'électricité (niveau de l'eau ou débit imposé pendant une partie de l'été).

A case in Europe is that River Rhine flowing through several countries: Switzerland, France and Germany, is the principal way for transporting goods to Switzerland (from Germany or France). Therefore there are important constraints to ensure its safety by means of a proper maintenance of these works, which have of course an incidence on the hydropower plants of the Rhine cascade.

Another case, this time in China, is the Yangtze River, which is also referred to as a 'golden waterway'. The Yangtze is the busiest river in the world in terms of goods transported. After the impoundment of the TGP, navigation conditions from Chongqing to the dam site improved significantly because of the elimination of all the shoals and rapids. The dredging depth for this section has been improved and some tributaries have become available for navigation. Moreover, re-regulation of the Gezhouba Project (GZB) downstream has reduced the impact of unsteady flows discharged from TGP.

2.3.6. Hydropower generation

As highlighted in the introduction, hydropower is the primary purpose of a significant proportion of large dams around the world. As world demand for energy increases, the demand for high dams built primarily for hydropower generation will undoubtedly grow.

Although hydropower is not a consumptive use of water, it does affect how reservoirs can or should be operated in many complex ways. Hydropower plants are concerned with reliability and safety of hydraulic structures and equipment, efficient usage of water resources (increasing power output) and the need to increase sales revenues. In the case of the cascade reservoirs and hydropower plants, integrated operation should be taken to maximize the efficiency of water use and sales revenues based on safety of hydraulic structures and equipment, while taking account of other water users' demands.

Some cases given in the bulletin do not introduce hydropower generation in detail, but almost all the relative cascades have the function of hydropower generation. Furthermore some hydropower cascades are mainly built for power generation such as the Itaipu Cascade in Parana River, which plays an important role in energy supply and promotes the development of economy significantly for Brazil, Paraguay and Argentina etc.

2.3.7. Recreation

Recreational users are interested in relatively stable water level in storage or downstream and over a certain time period or season (not all the year).

Canyoning, boating and kayaking are popular, especially during summertime; these activities must be taken into account by power producers for evident reasons of safety, which may result in time constraints, depending on the hydrological conditions of the river. In case of maintenance works on the plant, warnings should be given that conditions on the river will be degraded for a certain period of time. There are also needs for a certain amount of flow rate for boating activities, this is sometimes contractualized for sustaining flow rate during dry season or ensure a minimal activity during summertime (sustained development).

In France and Norway, recreational sites place demands on water in the lower reaches of dams; the Ariège valley in particular is used for many recreational activities. Canyoning, boating and kayaking are popular, especially during summertime, and some places are used as beaches, while an artificial water ski is installed on Lake Garabet during summer. All these activities result in constraints for the producer (the water level or flow rate is imposed for a period of time in summer).

2.4. GESTION DES CRUES ET DES ÉTIAGES

La prévention des catastrophes est un des principaux objectifs de la construction d'un barrage.

La crue est une catastrophe naturelle sévère. Beaucoup de barrage sont construits pour la protection contre les crues dans le besoin prioritaire; c'est le cas du barrage des Trois Gorges, qui protége 15 millions de personnes vivant au bord du Jingjiang, affluent de la rivière Yangtze.

L'étiage fait également partie des catastrophes naturelles. Avec le développement rapide de la population et de l'économie, le manque de ressource en eau est devenu de plus en plus sérieux, entrainant l'extension des zones arides et des effets des étiages. La lutte contre les étiages est devenue un souci des gouvernements, qui font des efforts pour limiter leur impact. Ainsi un guide de réponse aux étiages a été établi et mis en place en Corée.

2.4.1. Le management d'une crue

Le management des crues est fortement lié aux prévisions météorologiques et hydrologiques pour prévenir ou atténuer les dégâts, si possible en tirer des avantages. Ainsi le développement et l'application d'un outil de prévision est fondamental pour le management des crues.

2.4.1.1. Travail préparatoire

Les conditions locales peuvent influer sur les pratiques de surveillance des barrages et des ouvrages connexes, surtout avant les périodes de crue. En général, avant les crues saisonnières, on recommande les préparatifs suivants :

- Examiner l'état des ouvrages de retenue et en bordure du réservoir ;

- Entretenir les structures et les mécanismes d'évacuation des eaux ;

- Vérifier les lectures des instruments ;

- Vérifier le fonctionnement des vannes, des équipements de levage et des dispositifs de commande automatique ;

- Vérifier l'alimentation électrique d'urgence en cas de panne d'alimentation ;

- Former les équipes d'urgence à bien réagir en situation critique.

2.4.1.2. Abaissement du niveau du réservoir avant une crue

Pour certains réservoirs, il peut être nécessaire d'abaisser le niveau d'eau avant une saison de crue afin de réduire les risques d'inondation. Dans le cas d'un réservoir dont l'état géologique requiert des précautions, le débit de vidange doit être contrôlé de manière à assurer la stabilité des talus. Cette exigence peut aussi s'appliquer aux barrages en terre. On peut obtenir plus de détails dans les bulletins CIGB pertinents.

2.4. FLOOD AND DROUGHT MANAGEMENT

Disaster prevention and mitigation is one of the major purposes of dam construction.

Flood is a severe natural disaster. Many dams are built with primary purpose of flood control, such as the Three Gorges Dam to protect 15 million people lived along the Jingjiang reach of the Yangtze River.

Drought is among major natural disasters as well. As the rapid development of population and economy, water resources shortage has become increasingly serious, which lead directly to the expansion of arid area and aggravate drought degree. Drought has become a global concern and many efforts are made by governments for mitigation, such as "drought response guideline" are established and executed in Korea.

2.4.1. Flood management

Flood management should be highly linked with fine meteorological and hydrological forecasts to prevent or mitigate the damage, even produce benefits. Therefore, the development and application of forecast technology have a profound significance on flood management.

2.4.1.1. Preparatory work

Local conditions may influence practice regarding dam surveillance as well as their appurtenant structures in particular prior flood seasons. Before seasonal floods, it is generally recommended to do the following preparatory work:

- Examine the condition of water-retaining structures and the reservoir waterfront.

- Check the spilling structures if necessary.

- Test gauge indication, gates manipulating function, hoisting systems and automatic control equipments.

- Check stand-by power supply for emergencies, such as diesel engine.

- Accomplish regular maintenance and ensure the availability of generating units.

- Compile contingency plan and train emergency teams.

2.4.1.2. Reservoir drawdown before flooding

Certain reservoirs may need to be drawn down before a flood season to release capacity. For reservoirs with poor geological conditions, the water level drawdown rate should be controlled to ensure the stability of waterfront. Drawdown in earth dams may also need to be controlled. Details can be found in relevant ICOLD bulletins.

2.4.1.3. Opérations de gestion des crues

De nombreux barrages ont pour objectif d'assurer la sécurité de tronçons de rivière en aval, lors de crues majeures. Il est souhaitable d'avoir un système d'information et de prévision hydrométéorologiques afin d'assurer un contrôle du passage de la crue adéquat.

Dans le processus de laminage des débits de crue visant la sécurité des tronçons en aval, l'utilisation des groupes turbines-alternateurs est prioritaire dans la séquence d'évacuation, pour autant que leur fiabilité soit garantie. Lorsque le niveau d'eau dans la retenue approche le niveau haut de contrôle de la crue, une régulation par le réservoir est enclenchée pour équilibrer les débits entrants et sortants. Si le niveau d'eau ou le débit entrant continue d'augmenter, le processus du passage de la crue pour la sûreté du barrage est mis en place (assurer la sécurité de l'ouvrage tout en considérant les impacts en aval causés par le débit évacué).

Lorsque la pointe de crue a été atteinte, que le niveau maximal a été enregistré et que les apports se mettent à décliner, le niveau d'eau dans le réservoir sera abaissé aussi rapidement que possible jusqu'au niveau limite de gestion des crues, afin de pouvoir absorber la crue suivante sans nuire à la sécurité en amont et en aval.

Dans certaines rivières où les sédiments posent problème, l'exploitation du réservoir doit prévoir des opérations de chasse (évacuation des sédiment accumulés dans la retenue), afin de rétablir pour la prochaine crue la capacité de sédimentation du réservoir. Certaines procédures d'exploitation visent le nettoyage des ouvrages après une crue, surtout pour assurer que les équipements demeurent fonctionnels et ne soient pas menacés par les sédiments. On procède généralement par des crues artificielles, qui nettoient avec de l'eau propre les ouvrages évacuateurs de crue.

Beaucoup de cascades de réservoirs dans le monde ont une fonction de maîtrise et de gestion des crues, comme le barrage de Dez en Iran, la vallée du Rhône en Suisse et le complexe des Trois-Gorges en Chine. Tous ces aménagements ont des systèmes d'organisation complets, des règlements stricts et des modes opératoires efficaces pour la gestion du passage des crues.

2.4.2. Gestion des étiages

Les exploitants de barrage doivent établir un document de gestion pas à pas des étiages, il permet de préparer la situation d'étiage et de mettre en place les mesures nécessaires, comme un débit minimal restitué pendant la période d'étiage, ceci en référence au document de gestion des étiages. Un volume d'eau minimal doit être garanti pour assurer pendant l'étiage, que les inconvénients pour la population et les effets sur le corps social soient minimisés.

2.4.2.1. Règles de base pour l'approvisionnement en eau

Un guide de stockage journalier pour l'année suivante doit être établi pour le réservoir du barrage ; il définit les volumes à stocker. Pour une saison ordinaire, il est mis en place de manière flexible pour stocker l'eau à l'occasion des exploitations des barrages du bassin versant, en fonction des capacités de chaque barrage. Si le volume stocké à un moment donné est plus faible que la référence définie et qu'il est ainsi difficile de satisfaire la demande en eau à travers l'exploitation des retenues en cascade, les débits restitués à l'aval des barrages devront être réduits par étapes, après discussion avec les agences de l'eau concernées, pour éviter l'interruption future de la fourniture d'eau domestique et industrielle.

2.4.1.3. Flood control operation

Many stable dams are responsible for guarantying the safety of lower reaches when facing huge floods. Hydro-meteorological information, forecasting system and operating team should be adopted for a reasonable and scientific flood control regulation.

In the flood control regulation process for the safety of lower reaches, generators are prioritized in the water utilization sequence when their reliability is guaranteed. When the reservoir water level gets close to the flood control high level, regulation by spillway operation is carried out to balance inflow and outflow. If the water level or flow rate continues rising, the flood control regulation method for dam safety is adopted.

When peak flood flow has passed the dam with the highest water level been reached and the inflow is declining, the water level must be reduced as quickly as possible to flood control limit level to provide capacity for the next flood.

In some rivers with the sediment issue, sediment ejection should be taken into account in reservoir operation in order to reduce sedimentation during flood season, that is generally done by flushing through artificial flood.

Many cascade reservoirs across the world have flood control functions, such as Dez Dam in Iran, the Rhone River valley in Switzerland and the Three Gorges Project (TGP) in China. They all have complete organization structure, strict regulations and efficient operating modes for flood management.

2.4.2. Drought management

Dam operators should make a step-by-step plan called "drought response guideline" for preparing the drought situation in advance and carry out the necessary measures, such as the adjustment of water supply in time of event in accordance with the guideline. It must be required to ensure the proper water supply capacity for drought in order to minimize social and industrial effects.

2.4.2.1. Basic direction for dam water supply

Daily standard storage volume in guideline which stably supplies water from dams for the next one year should be set. For the ordinary season, it is flexibly implemented to supply water through conjunctive dam operation in river basin reflecting each dam's capacity. If the actual storage volume in particular time is lower than standard volume and it is hard to satisfy water demand in spite of conjunctive dam operation, water release from dams should be reduced by stages after the discussion with relevant agencies in order to prevent interruption of domestic or industrial water in the future.

2.4.2.2. Principe de base pour le guide de gestion des étiages

En général, il y a quatre types de fourniture d'eau pour un réservoir à buts multiples : usage domestique, industriel (y c. hydroélectricité), irrigation et maintenance de la rivière, la priorité étant donnée par le guide de stockage des apports. Ainsi la réponse aux étiages peut être définie par quatre niveaux de gestion en accord avec la demande en eau : vigilance, précaution, alerte et crise. Les niveaux de stockage par niveau garantissant 95% de la demande en eau, seront estimés en comparant le volume stocké en fonction des débits entrants prévus, avec la demande en eau.

Si la capacité de stockage du barrage approche le volume standard à stocker, une réduction du débit entrant dans le barrage sera mise en place en cohérence avec le guide de la réponse aux étiages. L'ordre de réduction est par ordre de priorité, la maintenance de la rivière, l'irrigation, les usages domestiques et industriels.

2.4.2.3. Plan d'action par étapes

Les plans d'action pour les exploitants de barrage et organismes associés sont établis en cohérence avec les niveaux de l'étiage. Pour le niveau vigilance, il faut ajuster le volume d'eau restitué aux valeurs contractuelles et assurer une surveillance en temps réel, tout en échangeant avec les différentes agences de l'eau concernées. Pour le niveau précaution, le débit d'eau pour la maintenance des rivières doit être réduit. Pour le niveau alerte, des débits pour l'irrigation doivent être réduits. Pour le niveau de crise, les débits d'eau domestique et industrielle doivent être réduits et des parades soutenables doivent être envisagées : construction d'un nouveau barrage, diversification des approvisionnements en eau, connexion de projets dans le bassin versant.

2.4.2.2. Basic principle in drought response guideline

Generally, there are four sorts of water supply from a multi-purposed dam, it would be classified into domestic, industrial, irrigational and river maintenance water supply according the priority. Therefore, drought response stage can be classified into 4 stages in accordance with water demand, namely notice, caution, alert and serious. Standard storage volume by stages which guarantees more than 95% safety for water supply would be estimated by comparing the storage volume reflecting the presumptive inflow with the water demand.

If the dam storage approaches the particular standard storage, the reduction of water supply will be implemented in accordance with drought response guideline. The order of reduction is primarily river maintenance, irrigational, domestic and industrial water.

2.4.2.3. Action plan in stages

Action plans for dam operators and related organizations are made in accordance with drought stage. For the notice stage, it is necessary to adjust water supply to contacted quantity and conduct real-time monitoring and progress sharing with relevant agencies. For the caution stage, river maintenance water should be reduced. For the alert stage, irrigational water should be reduced. For the serious stage, domestic and industrial water should be restricted and sustainable countermeasure including building a new dam, diversification of water sources and connection project cross the watershed can be developed.

3. LES CRITÈRES D'EXPLOITATION DE L'HYDROÉLECTRICITÉ

Les aménagements hydroélectriques convertissent l'énergie potentielle de l'eau en énergie électrique. Ils comprennent un ensemble de bâtiments, qui utilisent la puissance de l'eau pour la transformer en électricité, ainsi que des équipements pour la centrale hydroélectrique. Certains aménagements possèdent en plus des équipements pour permettre de passer les crues, réaliser les opérations d'irrigation, de navigation ou satisfaire à des contraintes environnementales.

Les aménagements hydroélectriques peuvent être classés de différentes manières. Premièrement selon l'origine de la ressource en eau, les centrales peuvent être différenciées en aménagements hydroélectriques classiques, en stations de pompage ou en aménagements marémoteurs. Deuxièmement selon la méthode d'exploitation du débit entrant, on peut les classer en aménagements au fil de l'eau ou en aménagements de stockage. Troisièmement selon la capacité d'ajustement des réservoirs, les centrales sont classées en exploitation pluriannuelle, annuelle, saisonnière, hebdomadaire ou journalière selon la taille du réservoir et des apports en eau.

Les aménagements hydroélectriques fournissent des services importants pour le réseau électrique, comme la régulation de fréquence, la régulation de la charge de pointe ou de base, la régulation de la tension et la constitution de réserves. Ils peuvent également réagir à toute variation de la charge et contribuer à la stabilité du réseau électrique par leur grande flexibilité et les services offerts. De plus c'est la seule possibilité de stocker de grandes quantités d'énergie à ce jour, en accumulant l'eau dans les retenues. En général pendant la saison humide les centrales peuvent produire plus en utilisant l'eau excédentaire, ce qui peut assurer la base de la courbe de consommation. Pendant la saison sèche, les centrales peuvent assurer la charge médiane et les pointes assurant ainsi un rôle important dans la régulation de la fréquence, la régulation des pointes et de la base et la constitution de réserves.

3.1. L'ÉQUILIBRE EN TEMPS RÉEL DE L'OFFRE ET DE LA DEMANDE EN PUISSANCE

La puissance électrique doit être fournie en bonne qualité pour répondre à la demande. La fourniture de puissance doit s'adapter à chaque instant pour suivre la courbe de consommation, qui fluctue en permanence et garantir ainsi une fréquence stable sur le réseau. Comme l'énergie de l'eau peut être stockée à travers les aménagements d'accumulation ou de pompage, elle participe au maintien de l'équilibre entre la consommation et la production, qui évolue en permanence.

Les changements de fréquence et de tension sur le réseau sont susceptibles d'affecter le fonctionnement des appareils électriques et la stabilité du réseau de transport de l'électricité. Lorsque les changements de fréquence et de tension dépassent un certain seuil, les protections des groupes peuvent déclencher, puis ensuite celles des autres groupes dans tout le système électrique. Ainsi cela peut conduire à un incident généralisé. En général si une défaillance se produit sur un groupe et si la reconstitution de la zone électrique n'est pas opérationnelle, cela conduit à une chute de l'efficacité de la production. C'est une des fonctions essentielles des aménagements hydrauliques que de contribuer à garantir la stabilité du réseau électrique.

3.1.1. Le mix énergétique et le rôle de l'hydroélectricité

Chaque pays a ses propres caractéristiques du mix énergétique, la fourniture d'électricité repose sur de multiples moyens, qu'il faut utiliser pour répondre à la demande des clients industriels et domestiques. Il est généralement composé d'énergie thermique nucléaire, thermique fossiles (charbon, fuel ou gaz), les énergies renouvelables et l'hydroélectricité qui a un rôle clé dans la production énergétique. On fait souvent la différence entre énergie de production de base (nucléaire, thermique) et celle qui sont dépendantes de la météorologie (renouvelables, hydraulique).

3. HYDROPOWER OPERATING CRITERIA

A hydropower station is an integrated facility which converts hydropower into electrical energy. It includes a series of buildings which use hydropower to produce electricity, as well as equipment for the hydropower station. Some hydropower plants not only include necessary buildings, but also have auxiliary buildings for the purposes of flood control, irrigation, navigation and ecological protection.

There are different types of Hydropower stations, including regular hydropower stations, pumped storage and tidal power stations according to the source of water, run-of-river and storage hydropower stations in the light of the adjustment method for natural runoff and multi-year, annual, seasonal, weekly, daily regulation hydropower stations depending on the capacity of the reservoir and inflow into the reservoir in accordance with the adjustment period of reservoir.

Hydropower stations undertake some major tasks in the power system, such as frequency regulation, peak and valley load regulation, voltage regulation and providing a reserve. They are able to rapidly react to load fluctuations and have great flexibility to offer several services for the stability of the power grid. Meanwhile, it is the only way of energy storage in large scale at present. Usually, during the wet season, water should be fully utilized to get more electrical energy, which could be the base load of the power system. During the dry season it should make up the intermediate load and peak load of a power system so that it could fully play an important role in frequency modulation, peak and valley load regulation, and emergency reserve.

3.1. REAL TIME POWER DEMAND-SUPPLY BALANCE

Electric power in good quality should be supplied stably to meet the demand. The capability of the power supply should change to meet consumption at any time and maintain a stable frequency because power demand changes constantly. As the water energy can be stored with storage hydropower stations or pumped storage hydropower stations, it helps the balance between supply and the ever-changing power demand

Changes of Voltage and frequency are likely to affect the operation of electrical equipment and the stability of a power system. When the changes exceed a certain level, they may activate protection of power generators resulting in one generator after another being tripped throughout the power system. Then, a large-scale power failure may be induced. In general, when a breakdown occurs, it will result in a drop of production efficiency if the electricity zone is not put back into operation. It is one of the main functions of HPP to keep the power grid stable.

3.1.1. Energy mix and the role of hydropower

Since each country has its own characteristics of energy mix, the power supply shall use multiple types of energy to fulfil industrial or domestic customers' needs. It is very often composed of nuclear, thermal (coal or gas fired) and renewable energy sources, with hydropower playing a key role in production. They are often further recognized as mass or meteorological dependant energy sources.

Généralement les énergies nucléaire, thermique au charbon et hydraulique au fil de l'eau, sont utilisées comme énergies de base pour satisfaire la demande. Les centrales au gaz liquéfié et au fuel, sont les sources principales d'énergie pour satisfaire la demande de semi-base. Enfin les centrales au gaz, de pompage-turbinage et hydroélectrique à accumulation, sont utilisées pour fournir l'énergie de pointe de consommation. Les coûts de production de ces différentes énergies sont bien sûr très différents.

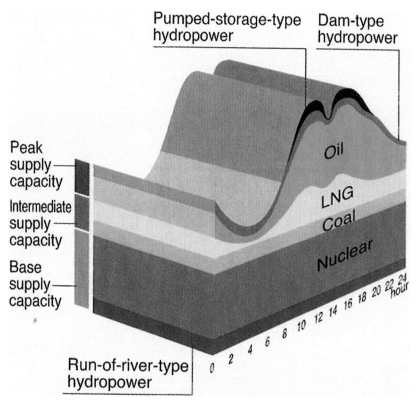

Figure 3.1
Bassin de la rivière Kiso (Japon) : situation des différentes énergies

Les politiques énergétiques de chaque pays varient selon leur propre contexte. Les zones électriques sont pilotées par plusieurs centres, le nombre dépendant de l'organisation de chaque pays concernant les zones électriques et les services associés. Il y a cinq zones régionales en Europe de l'Ouest, une en France, deux zones régionales en Chine et dix au Japon.

La production hydroélectrique est généralement réalisée loin des lieux de consommation. C'est pourquoi les réseaux électriques ont été développés depuis le début du 20ème siècle. Aujourd'hui les réseaux sont encore en évolution, même si de nouvelles technologies sont envisagées pour régler la question du transport de l'énergie, en utilisant des réseaux locaux et la technologie des réseaux intelligents (smart grids). Mais il y aura toujours des zones en manque d'énergie et les réseaux électriques devront être développés encore pour de nombreuses années. L'objectif principal est d'optimiser le bénéfice de la production électrique à travers le réseau.

Generally, nuclear, coal-fired thermal and run-of-river-type hydropower plants are used for base load. Liquid Natural Gas and oil thermal power plants are major power sources for intermediate load. Finally gas turbine, pumped-storage and dam-type hydropower plants are used for peak load. Production costs for these types of energy are of course very different.

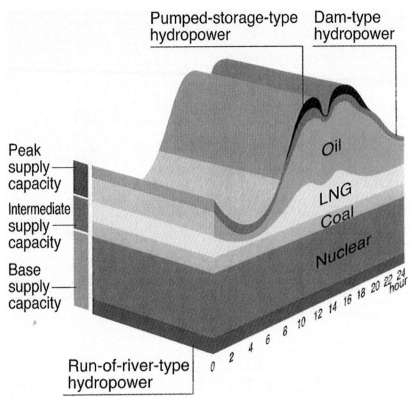

Fig. 3.1
From Kiso river (Japan), dispatching of power plants

Each country's energy policy varies according to its own situation. Electricity zones are managed by several centres; the number is dependent on the organisation of each country regarding electricity zones and service areas. There are five regional zones in Western Europe, one in France, two in China and ten in Japan.

Hydropower is generally produced far from places locations of consumption. This is why power grids had been developed since the beginning of the 20th century. Even if new technologies have been envisaged to solve the energy transmission issue such as local grids and smart grid technology, there will be always zones which lack electricity, and transmission grids will have to be developed for many years. The principal aim of the power grid is to optimize the benefit of electricity production.

L'énergie hydroélectrique contribue beaucoup à l'équilibre et à la stabilité du réseau électrique de puissance. Les centrales hydroélectriques équilibrent la demande et l'offre efficacement car elles sont capables de démarrer ou de s'arrêter en quelques minutes. Elles peuvent agir sur la puissance réactive du réseau en ajustant l'excitation du générateur, elles peuvent aussi contribuer efficacement aux problèmes qui surviennent lorsque l'on transporte sur le longues distances l'énergie (baisse de tension du réseau). Comme généralement les centrales hydroélectriques ont une plus grande habilité à répondre aux changements soudains de la demande, par rapport aux autres types de groupe de production, plusieurs centrales hydroélectriques sont équipées avec un contrôle automatique de la fréquence (CAF), le placement économique de la charge (PEC) et des dispositifs de régulation de vitesse pour les services systèmes. Lorsque qu'un incident de production se produit dans une centrale ou sur le réseau électrique, la fréquence du courant évoluera hors des limites admissibles. À ce moment, le CAF sera activé pour couper les charges peu importantes, avec l'objectif d'assurer la stabilité du réseau de puissance et la qualité opérationnelle du réseau.

Quelques pays sont riches en énergie hydroélectrique comme le Brésil, le Canada, la Norvège, la Chine, la Russie etc. Comme les ressources en eau sont généralement éloignées des lieux de consommation, l'exploitation hydroélectrique et la transmission de la puissance doivent avoir des caractéristiques de longues distances et de grande capacité. Ainsi la sûreté d'exploitation du réseau électrique de puissance peut être affectée par des phénomènes naturels comme le gel (la neige verglaçante), les typhons, les tremblements de terre, les incendies et la foudre. Parmi les exemples donnés, quelques situations météorologiques extrêmes peuvent concerner également la production hydroélectrique.

3.1.2. Dispatching de l'énergie

En fonction du mix énergétique et du dispatching de la consommation, chaque pays construit un réseau électrique de transport de la puissance, ils établissent un Centre de Dispatching Centralisé de l'Énergie (CDCE) et un ou plusieurs Centres Locaux de Dispatching et de Contrôle de la Charge (CLDCC). Ils sont responsables du dispatching de l'énergie, garantissent en temps réel l'équilibre entre production et consommation, pour maintenir une exploitation efficace économiquement et en toute sécurité.

Un CDCE supervise et contrôle l'équilibre entre la fourniture de puissance et la demande en énergie dans toute la zone de service électrique. Il supervise les différents CLDCC. Plusieurs CLDCC font une demande de production d'énergie à ceux qui exploitent et contrôlent les centrales, les sous-stations et les lignes de transport dans chaque zone de service électrique.

Quelques pays ne possèdent pas cette organisation, ou bien parce que leur chaine hydraulique a un seul objectif sans génération hydroélectrique, ou bien parce qu'il existe un acteur unique sur la chaîne hydraulique qui peut intégrer toutes les contraintes par lui-même, ou bien encore parce que la consommation et la production sont très proches géographiquement (il n'y a pas de lignes de transport). Mais le cas général est une exploitation intégrée de la chaine hydraulique, qui impose de partager les informations hydrauliques et de dispatching de l'énergie.

Pour exploiter le système électrique, les responsabilités entre les différents acteurs doivent être définies. Le réseau électrique connecte les zones de production et de consommation. Lorsque l'énergie hydroélectrique est produite, elle doit être intégrée localement autours de la zone de production ou transportée par le réseau électrique.

Les CLDCC ajustent la puissance produite à distance en actionnant les dispositifs CAF, PEC et régulateur de puissance pour les courtes et longues fluctuations de la demande en puissance à la minute ou à la seconde, comme montré à la Figure 3.2.

Hydropower stations contribute a lot to the balance and stability of power grids. Hydropower balances supply and demand efficiently with the benefit of being able to start up and stop within minutes. It meets the reactive power in a system by adjusting the excitation of generator, which also effectively alleviates problems coming from long distance energy transmission (voltage reduction). As generating units of hydropower plants have a better ability to respond to sudden changes in demand than other types of unit, several hydropower plants are equipped with Automatic Frequency Control (AFC), Economic Load Dispatching (ELD), and Governor-free (GF) devices for ancillary service in service areas. When there is an incident in the power supply or the power grid, the frequency would go out of limit. At this time, AFC would be activated to cut some unimportant loads, in order to ensure the stability of the power grid and operational quality of power.

Some countries are rich in water resources, such as Brazil, Canada, China, Norway and Russia etc. Since water resources are far from the places of consumption, hydropower exploitation and power transmission may have the characteristic of long-distance and high capacity. Therefore, the safe operation of a power transmission network may be affected by natural phenomena such as freezing, typhoons, earthquakes, fires, and thunder storms. Among the above mentioned factors, some extreme meteorological situations may also be the concerns of the hydropower production.

3.1.2. Power dispatching

Each country constructs a power grid network according to its energy mix and distribution of consumption and establishes a Central Load Dispatching Centre (CLDC) and one or several Local Load Dispatching and Control Centres (LLDC). They are responsible for power dispatching and guarantee the balance between energy supply and demand in real time to achieve safe and efficient operation.

A CLDC supervises and controls the balance between the supply and demand of power in the entire service area. It supervises different LDCC. LDCC make a request for generation operation to those who operate and control plants, substations, and transmission lines in each service area.

Some countries do not have all these divisions, either because their cascades have a unique purpose that does not include hydropower generation, or because there is a unique actor on the cascade who can manage all the constraints by himself, or because consumption and production areas are very close (no transmission lines). But in general, there is integrated operation of HPP that needs sharing of hydrological and dispatching information among them.

To manage the power system, responsibilities of these numerous actors must be defined. The power grids connect production and consumption areas. When hydropower is produced, it should be integrated locally around the production site before transmitted to the grid.

The CLDC regulate the power output remotely by controlling the AFC, ELD and GF devices in long and short term, responding to periodical load fluctuations by minute or second, as shown in Figure 3.2.

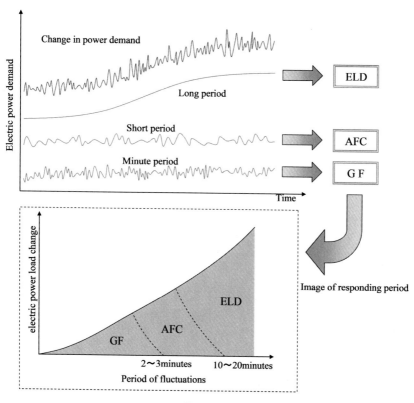

Figure 3.2
Services au système électrique en fonction des types de fluctuation de la demande

En prenant pour exemple le réseau japonais, on trouve dans le graphique ci-dessous la chaîne de commande, la structure de partage des données, ainsi que la répartition des différentes tâches.

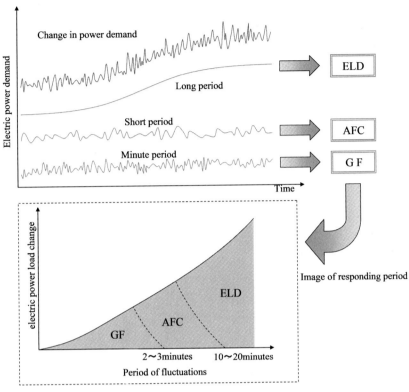

Fig. 3.2
Ancillary service in accordance with types of fluctuation in demand

Take a Japanese power utility for example, the related units' chain command and data sharing structure as well as the major duties are shown below.

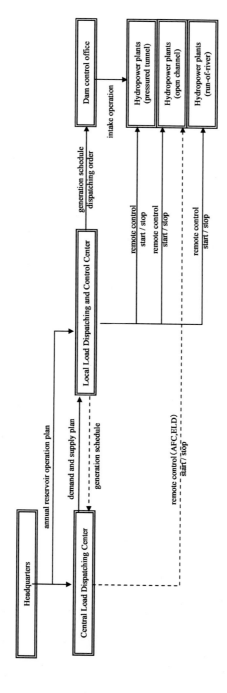

Figure 3.3
Chaine de commande pour l'exploitation des centrales

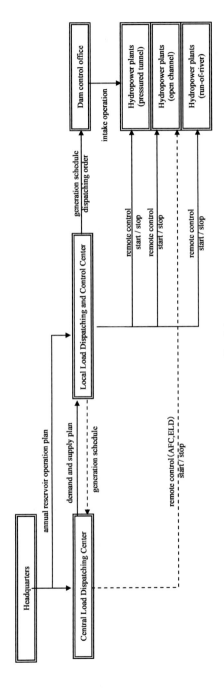

Fig. 3.3
Chain of command for operating units

3.2. MANAGEMENT DES RETENUES ET CENTRALES HYDROÉLECTRIQUES

3.2.1. *Management du plan de production*

L'Exploitant du Système de Transport de la production (EST) est responsable de la qualité et de la stabilité du réseau électrique. En accord avec les prévisions de demande et de production, l'EST transmet les instructions d'exploitation aux producteurs, qui ont leur propre organisation pour programmer et contrôler leur production. Les producteurs hydroélectriques sont responsables de la qualité et du contrôle de la fourniture de tous les aménagements le long de la cascade.

L'EST étudie et prévoit la demande annuelle en énergie et établit un plan de production pour chaque producteur selon la répartition des zones de production électrique ou des zones de services au système électrique.

Sur la base d'une exploitation en sécurité du système électrique et des centrales, les producteurs proposeront à l'EST des plans de production, qui prennent en compte les objectifs, les critères d'exploitation (comme la puissance maximale disponible, la disponibilité de régulation de la retenue, les priorités d'ordre d'exploitation, le plan de maintenance annuel, etc.), ainsi que les principes pertinents pour l'exploitation des retenues et des centrales hydroélectriques. La procédure d'élaboration de ce plan comprend :

- Une collecte des données hydraulique, et météorologiques, des prévisions (sur le court terme, le moyen terme et le long terme) en s'assurant de la précision de ces prévisions.

- L'utilisation des capacités en eau pour contrôler et gérer les retenues en même temps que la production des centrales hydroélectriques ; dans le même temps il faut prendre en considération les demandes concernant le passage des crues, la navigation, l'irrigation, l'approvisionnement en eau potable et les activités récréatives.

- Sur la base de l'analyse des facteurs d'influence (comme les prévisions de débit des rivières, le plan d'exploitation de la retenue, etc.), des capacités de production, l'exploitant établit un plan de production (qui inclue la fourniture en énergie et en puissance des groupes et un plan de maintenance) pour les différentes échéances de temps (journée, semaine, mois, année).

- Une communication et une coordination avec l'EST sont établies pendant l'élaboration du plan, ensuite soumis à l'EST. Dans les pays où une bourse de l'énergie est en place, les programmes journaliers sont établis par les offres des producteurs. Lorsque le programme final est établi par l'exploitant du réseau électrique, la production réelle doit suivre ces programmes. Si nécessaire, les programmes peuvent être modifiés en infra journalier par un système d'offre.

Finalement le plan de production est soumis à l'EST, qui sera mis en œuvre après validation. Dans les pays où il n'existe pas de bourse de l'énergie, il est important de conserver la communication et la coordination entre les producteurs et l'EST pendant l'établissement du plan de production. Dans les pays où existe une bourse de l'énergie, les programmes journaliers sont établis par appel d'offre. Après réception du programme de production finalisé par l'exploitant du système électrique (EST). La production réelle doit suivre le programme, qui peut être changé par appel d'offre à travers une procédure infra-journalière si nécessaire.

3.2. RESERVOIR AND HYDROPOWER STATION MANAGEMENT

3.2.1. Power Production plan management

The Transmission System Operator (TSO) or CLDC which is mentioned above is responsible for power quality and the stability of the power grid. In accordance with the supply and demand plan, the TSO directs generating instructions to producers who have their own organization to dispatch and control production, as different producers are responsible for controlling all the hydropower plants on their cascade.

The TSO studies or forecasts the annual power demand and establishes a production plan for all producers, according to power system division or services areas.

On the premise of secure operation of the power grid and plants, producers shall propose production plans to the TSO, based on the targets, parameters (such as maximum power output, available capacity of regulating pond, priority order in operation, annual outage plan, etc.) and relevant principles of reservoirs and hydropower stations. The procedure for drawing up the plan includes:

- Collecting hydrological and meteorological data, making forecasts of the electrical demand (including short-term, mid-term and long-term) and ensuring the accuracy of the forecasts.

- Utilizing beneficial capacity to balance reservoir storage and control electricity production; in the meantime, giving consideration to the demands of flood control, navigation, irrigation, water supply and recreational activities.

- Based on the analysis of influential factors of power generation capacity, (such as river flow forecasts, reservoir operation plans, etc.), establishing a production plan (includes power generation plan and maintenance plan) for different periods (day/week/month/year).

- Keeping communication and coordination with the TSO during the drawing up of the production plan and submitting the finished plan to the TSO.

Finally, the production plans would be submitted to the TSO, which would then be confirmed and carried out. In countries where there is no energy pool, it is important to maintain communication and coordination with the TSO during the drawing up of the production plan. In countries where an energy pool is in place, the daily program is determined by bidding. After receiving the final daily program from the operator, actual output should follow the programs, which could be changed by bidding on the intra-day market if necessary.

Certains pays attachent une grande importance à l'utilisation de la valeur de l'eau. Une division spécifique au sein d'EDF en France est destinée à optimiser le placement des groupes de production. Les plans de production des centrales hydroélectriques sont optimisés en utilisant la notion de 'valeur de l'eau'. Aux Etats-Unis, la TVA Tennessee Valley Authority et son River Forecast Center (RFC) est responsable de fournir des prévisions hydrauliques et météorologiques, de produire des programmes horaires de production des centrales hydroélectrique de la TVA, de fournir les lâchers d'eau nécessaires pour les autres besoins en eau sur les retenues de la TVA, avec pour objectif de maximiser la 'valeur de l'eau'. En Russie, les programmes sont établis à partir d'algorithmes basés sur le régime des eaux, sur l'état des lignes d'évacuation de l'énergie et en prenant également en compte les autres limitations pour satisfaire tous les besoins. La cible de ce plan de production est également d'optimiser la 'valeur de l'eau'.

3.2.2. *Le management des centrales hydroélectriques*

L'EST et les producteurs doivent prendre leur responsabilité, en utilisant les systèmes de supervision et de contrôle pour surveiller et piloter le processus de production de l'énergie électrique, assurant ainsi une exploitation économique et sûre du système électrique.

3.2.2.1. La régulation en temps réel de la production des aménagements hydroélectrique

Une offre de production est faite chaque jour à l'EST par les producteurs, incluant les capacités de services système ; l'EST utilise ces offres pour satisfaire ses besoins et garantir la stabilité du réseau électrique.

Les producteurs mettent en place les moyens en personnels et en travaux programmés, afin de mettre en œuvre le plan de production, la surveillance en exploitation des équipements et la mise en œuvre des consignes de production. Chaque heure, les producteurs ont la possibilité de sortir des groupes en fonction soit d'incidents, soit d'évolutions dans les paramètres de prévision. Si les écarts ne sont pas corrigés, le coût est répercuté par le régulateur du réseau sur les producteurs à l'origine de ces écarts.

Pour les producteurs qui ont construit plusieurs aménagements dans le même bassin hydrographique, ils implantent généralement un centre de dispatching et de contrôle centralisé pour leurs centrales hydroélectriques, avec l'objectif de maximiser le bénéfice des aménagements en cascade, en prenant également en compte les demandes de gestion des crues, de navigation, d'irrigation et de préservation de l'écosystème. Pour ces raisons, les producteurs mettent en place des centres de contrôle des aménagements responsables de l'optimisation du fonctionnement des centrales, des opérations de gestion de crue sur les réservoirs des aménagements en cascade, de recevoir et mettre en place les ordres de dispatching des groupes de production. Ils sont également responsables d'établir, de soumettre puis de mettre en place les programmes de production comme les plans de régulation des réservoirs de stockage.

3.2.2.2. Gestion des conflits entre production électrique et des autres besoins en eau

Pour les aménagements hydroélectriques les sujets importants sont la fiabilité des structures et des équipements, l'utilisation efficace de l'eau ainsi que les revenus de la vente de l'électricité. Ainsi il est nécessaire de contrôler et gérer le débit dans la rivière pour maximiser l'énergie produite en satisfaisant les autres demandes de ressources en eau. Bien que la production hydroélectrique ne consomme pas d'eau, il peut y avoir des conflits entre la production électrique et les autres utilisations de l'eau :

Some countries attach great importance to the usage of the 'value of water'. A specific division has been established in France at EDF to optimize the placement of production groups. The production plans of the hydropower stations are therefore optimized by analysing the 'value of water'. In the USA, the TVA River Forecast Centre (RFC) is responsible for issuing forecasts of hydrological and meteorological data, providing hourly generation schedules for TVA hydroelectric projects and scheduling water releases at TVA dams for other water demands with the goal of maximizing the 'value of water'. In Russia, production plans are made strictly according to certain algorithms based on water regimes and different dispatching lines and zones after taking other limitations into account. The production plan's optimal goal is also maximizing the 'value of water'.

3.2.2. Hydropower management

The TSO and producers should take due responsibility, utilizing computerized supervisory control systems to monitor and control electricity production process, thereby ensuring secure and economic operation of the power system.

3.2.2.1. Real-time production regulation of hydropower stations

A production offer is submitted each day to the TSO by producers, including ancillary services capability; the TSO takes up these offers and makes production plan according to the needs of the power grid on a stable and economical basis.

Producers allocate full-time staff for production plans' implementation, equipment operation surveillance and execution of dispatching orders. Each hour, producers may shift units, depending on incidents or other variations in forecasting values. If generation output does not follow the plan strictly in certain period, the producer will be charged or even punished.

For producers that have developed several hydropower stations in the same river basin, they usually implement centralized dispatch and control of their hydropower plants so as to maximize the comprehensive benefit of the cascade and in the meantime give consideration to the demands of flood control, navigation, irrigation and ecosystems. For these reasons, the producers set up centralized control centres, which are responsible for the optimised generation operation and flood control operations of their cascade reservoirs, to receive and implement the dispatch orders. They are also in charge of editing, submitting and implementing power generation plans as well as implementation schemes for storage regulation.

3.2.2.2. Contradictions between hydropower production and other water user requirements

For hydropower stations, there are important matters such as the reliability of the hydraulic structures and devices, efficient utilization of water, as well as electricity sales revenue. Therefore, it is necessary to control and manage the river flow in order to maximize power output on the basis of meeting other demands for water resources. Although hydropower generation does not consume water, there may be contradictories between hydropower generation and other water use demands:

1. Conflit entre production électrique et gestion du passage de la crue. Pour les réservoirs avec une capacité de rétention des crues, le niveau d'eau pendant la saison des crues devrait être maintenu en deçà du niveau pour lequel la crue conduirait à un déversement par les vannes. C'est en contradiction avec la demande de fourniture d'énergie qui demanderait à conserver l'eau dans la retenue. Mais avec le développement des technologies d'exploitation et l'amélioration des méthodes de prévision hydrologiques, le niveau limite pour une crue donnée peut être adapté en temps réel pour réduire le débit déversé autant que possible.

2. Conflit entre production hydraulique et navigation. D'un coté le réservoir fournit de l'eau aux biefs inférieurs pour répondre à la demande de navigation pendant la saison sèche, ce qui abaisse l'efficacité de production d'énergie avec un niveau plus bas de la retenue. D'un autre côté, les aménagements hydroélectriques du fait de leur flexibilité, jouent un rôle important pour le système électrique dans la régulation des pointes et de la base. Ainsi le débit sortant n'est pas stable, ce qui est en contradiction avec les besoins de la navigation en aval, qui nécessite des niveaux stabilisés.

3. Conflits entre production hydraulique et tous les autres besoins en eau en aval. Afin de satisfaire les besoins en eau des industriels et des particuliers, ainsi que les demandes en période d'étiage ou pour limiter les intrusions d'eau saline, certaines retenues doivent baisser leur niveau d'eau, réduisant ainsi leur efficacité à fournir de l'énergie.

Pour un réservoir à buts multiples, chaque aspect de l'opération globale a sa propre programmation d'exploitation, ces programmes sont reliés et parfois contradictoires. Il est nécessaire de prendre en considération tous les besoins en eau et de maintenir un équilibre entre eux, de comparer différentes stratégies d'exploitation, puis de choisir la 'meilleure', qui maximise la 'valeur de l'eau'.

3.2.2.3. Prise en compte des opérations de maintenance

Les programmes de maintenance sont anticipés au moins une année à l'avance, les opérations importantes doivent être intégrées en début d'année pour les années suivantes. Pour les centrales hydroélectriques, les opérations conjointes devraient être prise en compte, spécialement pour les aménagements en cascade.

La maintenance régulière doit être programmée raisonnablement et ajustée en temps réel pour s'accorder avec l'état récent des centrales et des besoins du réseau, ceci non seulement pour garantir la fiabilité des groupes de production et des autres équipements, mais aussi pour réduire les pertes par déversement et ainsi augmenter le taux d'utilisation de l'eau.

Le dispatching et le contrôle à distance des équipements a été mis en place dans beaucoup de pays, utilisant des technologies avancées et des organisations adaptées, cependant du personnel spécialisé dans la maintenance des appareils de contrôle particulièrement dans les grandes installations hydroélectriques devrait être mis en place. Ce personnel devrait être en charge de la maintenance des équipements d'évacuation des eaux, de la confirmation de l'état des équipements et des actions en cas d'urgence pour répondre aux demandes du dispatching et du contrôle et de l'exploitation des centrales hydroélectriques.

3.2.2.4. Dispatching et contrôle

Les personnels d'exploitation sont principalement en charge du temps réel pour la gestion du ou des réservoirs et de la production hydroélectrique ; ils conduisent la gestion et le contrôle des retenues, des structures d'évacuation des eaux, des équipements électriques et mécaniques, des équipements auxiliaires, ainsi que l'introduction des programmes de production à travers un système automatisé fiable. De plus, ils jouent un rôle essentiel dans les réponses à apporter en cas d'urgence, pour garantir une exploitation stabilisée et sûre des aménagements hydroélectriques.

1. The contradictions between power generation and flood control. For reservoirs with flood control capability, water levels during flood seasons should generally be maintained at the flood control limitation level while the surplus water should be spilt through the gates. It contradicts with the demands of power generation, in which the water should be kept in the reservoir. But with the development of operating technologies and increased accuracy of hydrological forecasting, the flood control limitation level can be adjusted dynamically to reduce the discharged flow as far as possible.

2. The contradictions between power generation and navigation. On the one hand, the reservoir generally replenishes water to the lower reaches to meet the demands of navigation during drought periods, leaving the reservoir with reduced power generating efficiency and a lower water level. On the other hand, hydropower stations play a key role in peak and valley load regulation of the grid due to their flexibility, thus the discharged flow is unstable so that it is contradictory with downstream navigation, which requires a stable water level.

3. The contradictions between power generation and the comprehensive utilization of water in the lower reaches. In order to meet industrial and residential water demands requirements, as well as demands to relieve drought and salt tides, some reservoirs may lower the water level during drought periods, thus reducing power generating efficiency.

For a multi-target reservoir, each aspect of comprehensive operation has its unique operation schedule; these schedules are related but sometimes contradictory. It is necessary to take all the water demands into consideration and keep a balance between them, comparing different management strategies and selecting the best one in terms of maximising the "value of water".

3.2.2.3. Maintenance arrangement

Maintenance programs are anticipated at least one year in advance, and important operations must be integrated at the beginning of the year for the following year. For hydropower stations, joint operations should be taken into consideration, especially for units on a river cascade.

Regular maintenance should be reasonably arranged and adjusted dynamically according to latest situations, not only to ensure the reliability of units and other equipment, but also to reduce water spillage and enhance water utilization rates.

Remote dispatch and control of field equipment has mostly been realized in many countries through advanced technologies and management, yet a certain number of specialized staff must be assigned to field device maintenance in some large or important HPPs. They should be in charge of the maintenance of discharge facilities, confirmation of equipment status and emergency responses to meet the requirements of dispatching, control and operation of hydropower stations.

3.2.2.4. Dispatch and control

Staffs on duty are mainly in charge of real-time reservoir dispatch and power production, conducting real-time monitoring and control of reservoir operations, spillway structures, electrical equipment, mechanical equipment and auxiliary devices, as well as implementing power generation plans through reliable automation systems. In addition, they play a key role in emergency responses to ensure stable operation of hydropower stations.

Quelques pays développés ont installé des systèmes de démarrage et d'arrêt à distance pour la plupart de leurs groupes de production. Au Japon 13 barrages et 33 aménagements hydroélectriques sur la rivière Kiso sont tous pilotés à distance. En France les aménagements hydroélectriques sont généralement sans personnel, sauf certaines ou les équipes d'exploitation et de maintenance sont nécessaires sur place. De façon générale, un point central a la charge de plusieurs aménagements. Tous les grands aménagements hydroélectriques sont commandés à distance. Les autres pays disposent de programmes semi-automatiques, qui doivent être validés par le producteur parce qu'il est responsable de la sûreté de ses installations, en particulier en ce qui concerne les débits relâchés dans la rivière. Les solutions apportées varient selon les caractéristiques du bassin versant et le rôle des aménagements hydroélectrique dans l'exploitation du système électrique.

En France à EDF, l'un des grands producteurs d'électricité, il y a quatre centres de placement de l'énergie, qui exploitent seulement 20% des aménagements hydroélectriques, mais leur production représente 80% de l'énergie hydroélectrique produite. Le reste des aménagements hydroélectriques est exploité par les producteurs, selon des procédures semi-automatiques. En Chine, les aménagements de Xiluodu et Xiangjiaba sur la partie basse de la rivière Jinsha ont mis en place un système d'exploitation et de contrôle à distance de la production des groupes et des écluses.

3.3. LE RÔLE DES STATIONS DE POMPAGE

Les stations de pompage sont équipées avec des réservoirs de stockage inférieur et supérieur. L'eau du réservoir inférieur est pompée vers le réservoir supérieur en utilisant les groupes de pompage-turbinage la nuit, lorsque la consommation en puissance est basse et donc l'énergie bon marché. Elle est restituée en journée pendant le pic de consommation en puissance, lorsque l'énergie est chère, en turbinant l'eau du réservoir supérieur vers le réservoir inférieur. Les stations de pompage et de stockage minimisent ainsi la production des centrales thermiques, dont le coût de production est élevé et émettent de grande quantité de CO_2. Bien qu'il y ait des pertes d'énergie dans ces transferts d'eau, cela reste une forme d'énergie propre, meilleure que celle des turbines à gaz (production de CO_2) et qui joue de plus un rôle clé dans la régulation du système électrique.

Ainsi la puissance électrique peut être stockée en pompant de l'eau d'une retenue inférieure vers une retenue supérieure. Jusqu'à présent c'est la seule manière de stocker en masse de l'énergie, même si les batteries se développent beaucoup actuellement, le réservoir de stockage demeure compétitif. Les générateurs sont démarrés et stoppés plus rapidement que les centrales thermiques ou d'autres formes d'énergie (exceptées les turbines à gaz, mais qui ne sont pas respectueuses de l'environnement). Les stations de pompage sont ainsi comme une énorme batterie qui peut servir d'alternative lorsqu'une centrale hydroélectrique de grande capacité est arrêtée en urgence. Elles jouent un rôle essentiel dans l'équilibre entre la production et la consommation.

Cependant environ 30% de l'énergie est perdue dans un cycle de pompage à cause principalement de l'efficacité des groupes de pompage turbinage et des pertes dans les circuits hydrauliques. Cela peut apparaître négatif, mais cet aspect est compensé par la différence de prix de l'énergie entre les périodes de faible (pompage) et forte consommation (turbinage), sans compter d'autres avantages comme la capacité à stocker l'énergie, etc.

Plusieurs pays ont récemment fait évoluer leur politique énergétique, en privilégiant les énergies renouvelables, réduisant ainsi la dépendance de la production d'énergie basée sur les combustibles fossiles. Comme elles sont météo-dépendantes, et ont des difficultés à être stockées, l'utilisation des énergies renouvelables comme le solaire ou le vent en grande quantité a un impact sur la variabilité de la production, qui doit être contrecarrée par d'autres sources d'énergie, en particulier l'hydroélectricité de par ses qualités spécifiques. La capacité d'ajustement de l'offre et de la demande en électricité du système électrique global devra être améliorée de manière importante dans le futur.

Some developed countries have implemented remote start and stop for most of their generating units. In Japan, 13 dams and 33 hydropower plants on the Kiso River system and all of the power plants are unmanned. In France, HPPs are generally unmanned, except those by which maintenance and operating teams are needed on site. Generally speaking, a central point is in charge of several HPPs. And all large HPPs are remote controlled. Other countries have semi-automatic programs that have to be confirmed by the producer, who is responsible for the safety of its plants (regarding runoff in the river in particular). Solutions vary depending on the watershed characteristics of the hydropower stations and their role in the power grid.

In France at EDF, one of the main producers, there are four dispatching centres that operate about 20% of the HPPs but represent more than 80% of the hydroelectric energy produced. The rest of the HPPs are operated by producers, according to semi-automatic procedures. In China, Xiluodu and Xiangjiaba hydropower plants on the lower reaches of Jinsha River have realized remote centralized dispatch and control of giant generation units as well as sluice gates.

3.3. THE ROLE OF PUMPED STORAGE POWER PLANTS

Pumped storage power plants are equipped with upper and lower storage reservoirs. The water in the lower storage reservoir is pumped up to the upper storage reservoir using pump turbines and low-cost electric power during the night when power consumption is low. Power is generated during the day at peak power consumption, when energy prices are high by releasing water from the upper storage reservoir into the lower storage reservoir. Thus, pumped storage power plants minimize output from thermal power plants that are costly and emit large quantities of CO_2. Although there is a loss in energy efficiency across the energy transfers, it is still a kind of clean energy that is better than gas turbines and furthermore plays a key role in peak and valley regulation.

Power can be stored as potential energy by pumping water into the upper storage reservoir. This is currently the only possibility for mass storage of energy, even if batteries are developing a lot these times, reservoir storage remains competitive. The generators are activated and stopped more quickly than thermal or other power sources (except gas turbines, which are not environmentally friendly). Pumped storage power plants therefore also serve as a large battery alternative in cases where a large-capacity power plant is brought to an emergency stop. They play key role in balance of supply and demand.

Approximately 30% of energy, however, is lost in each cycle from power generation to water pumping because of such factors as the mechanical inefficiency of generators and friction loss in the headrace channel. This seems to be a little bit negative but is compensated by the interest of the price differences between low and high consumption periods, apart from other advantages of storage of energy and so on.

Several countries have recently undergone a policy shift to using power generation from renewable sources with reduced dependence on fossil fuel generation. Since they are more meteorological dependant and still have difficulties to be stored, using renewable power sources such as solar and wind power in large quantities has an impact on the variability of the production that has to be compensated by other sources of energy, particularly hydropower because of its specific qualities. The supply and demand adjustment capacity of the entire power system needs to be greatly enhanced in the future.

La construction de nombreuses stations de pompage est programmée dans les années à venir dans plusieurs pays européens, où les sources d'énergie renouvelable représentent une part importante du mix énergétique (principalement l'Allemagne, la Suisse et l'Autriche). L'objectif est d'améliorer la capacité d'ajustement entre l'offre et la demande. Les stations de pompage seront construites d'ici 2020 sur 60 sites à travers l'Europe, fournissant une capacité de 27 000 MW, soit une augmentation de 60% par rapport à la puissance disponible des stations de pompage aujourd'hui : 45 000MW en 2015.

Le système de station de pompage à vitesse variable est exploité sur les stations de pompage où la vitesse de rotation de la turbine peut être ajustée librement lorsque l'eau est remontée de la retenue inférieure à la retenue supérieure. C'est un nouveau moyen d'ajuster la puissance produite ou consommée par la station de pompage. Une centrale de pompage conventionnelle ne pompe qu'à une vitesse fixée et la puissance produite est aussi fixée, la centrale n'est pas adaptée pour s'ajuster entre offre et demande. Le système de station de pompage à vitesse variable permet l'ajustement en puissance pendant la phase de pompage, il contribue ainsi grandement à l'équilibre offre-demande, particulièrement la nuit lorsque la demande est faible et les capacités d'ajustement faibles pour les centrales thermiques dans certains réseaux électriques.

3.4. COORDINATION AVEC LES AUTRES SOURCES D'ÉNERGIE

Avec le développement de la société, les dégradations environnementales deviennent un problème de plus en plus sérieux. Pour cette raison, beaucoup de pays veulent augmenter la part d'énergie renouvelable comme le solaire ou l'éolien dans leur mix énergétique.

De nos jours, de nombreux pays ont adopté des politiques visant à renforcer le développement des énergies renouvelables. Ces politiques peuvent être divisées en deux systèmes : à prix fixés ou à quantités fixées. Pour les systèmes à prix fixés, le prix est déterminé par le gouvernement et la quantité est déterminée par le marché. Le second système fixe des quantités; ce type de système est aussi dénommés système à quota d'énergie renouvelable. Dans ce système, le gouvernement définit les quantités d'énergie renouvelable par quota et les prix sont déterminés par le marché. Ces politiques ont principalement favorisé les énergies solaire et éolienne, pour les rendre compétitives par rapport aux formes d'énergie traditionnelle sur le marché.

Cependant, la production d'énergie éolienne et solaire photovoltaïque est très sensible aux conditions météorologiques, de sorte que les fluctuations qu'elles peuvent provoquer dans le mix énergétique mondial sont de plus en plus importantes, ce qui constitue un réel problème pour les réseaux électriques.

L'énergie éolienne comme une alternative aux énergies fossiles est abondante, renouvelable, largement répartie, propre, ne produit aucun gaz à effet de serre. Comme l'énergie du vent est aléatoire et intermittente, les fermes éoliennes ne peuvent pas produire une puissance stable. À cause de cette instabilité et discontinuité, le système électrique doit faire face à un problème important lorsque que la production d'énergie éolienne est transmise au réseau électrique. L'exploitation de l'énergie éolienne est aussi limitée par la demande d'énergie. Mais les fermes éoliennes pourraient bénéficier des caractéristiques des centrales hydroélectriques, en les complétant par des centrales de pompage, qui ont une réponse et une montée en puissance rapides et des capacités de stockage de l'énergie.

L'énergie photovoltaïque a longtemps été vue comme une technologie d'énergie propre et durable, elle répartit sur la planète une énergie renouvelable en grande quantité : le soleil. Mais c'est aussi une énergie sensible à la météorologie, qui n'est donc pas stable. Les centrales nucléaires ont les coûts de fonctionnement les plus faibles et peu d'impact environnemental, mais le combustible utilisé génère des risques pour l'environnement. Lorsqu'une fuite du combustible nucléaire se produit, l'environnement immédiat sera dévasté. Comme la capacité d'un aménagement nucléaire est très grande, l'impact d'un arrêt inopiné ou d'un arrêt pour maintenance sera fort sur le réseau électrique. Ainsi la construction de grands aménagements hydroélectriques ou de station de pompage, pourrait aider à compenser ces variations de charges et ainsi améliorer l'efficacité des centrales nucléaire.

Numerous pumped storage power plants are scheduled to be constructed in some European countries where renewable energy sources have a large share of the market and in adjacent countries (mainly Germany, Austria and Switzerland), with a view to enhancing supply and demand adjustment capacity. They will be built at 60 locations throughout Europe by 2020, providing a capacity of 27 GW, increased by 60% on the present 45 GW provided by existing pumped storage power plants.

The variable-speed pumped storage power generation system is being operated at pumped storage power plants under which the rotational speed of pump turbines can be set freely when water is pumped from the lower storage reservoir to upper storage reservoir as a new means of adjusting output. The conventional type of plant could be pumped only at a fixed speed and output was not fit for adjusting to power supply and demand. The variable-speed pumped storage power generation system enables the adjustment of power during the pumping operation and makes great contributions to supply and demand adjustment during the midnight period when demand is low and the adjustment capacity from thermal power plants is insufficient in some power grids.

3.4. COORDINATION WITH OTHER RENEWABLE SOURCES

As society develops, environmental degradation becomes an ever more serious problem. For this reason, many countries want to increase the proportion of renewable energy production, such as wind and photovoltaic energy in their energy mix.

Nowadays, many countries have made policies to enhance the development of renewable energies. These policies can be divided into two systems, fixed-price systems and fixed-quantity systems. In fixed-price systems, a fixed price is formulated by the government and the quantity is determined by the market. The alternative is fixed-quantity systems, which are also called renewable-quota systems. In this type of system, the government restricts the amount of renewable energy production, but the price is determined by the market. These policies largely benefit wind and solar production plants and are implemented in order to ensure that renewable energy can compete with traditional energy in the market.

However, wind and solar photovoltaic production are very sensitive to weather conditions, so the fluctuations they may cause in the global mix of energy are of increasing importance, which is a real issue for power grids.

Wind energy, as an alternative to fossil fuels, is plentiful, renewable, widely distributed, clean, and produces no greenhouse gas emissions during operation. Since wind energy is random and intermittent, wind farms cannot provide stable power. Due to the instability and discontinuity, power system would face a big challenge when wind energy is transmitted to the grid. The operation of wind energy generators is also be limited by load demand in the power system. Wind energy plants can benefit from the characteristics of hydropower units, including pumped storage power units, which have the characteristics of quick reaction, fast ramp-up and can store energy.

Power generation from photovoltaic is a clean and sustainable energy which draws upon the planet's most plentiful and widely distributed renewable energy source-the sun. However, it is also a weather-based energy that is not stable. Nuclear power plants have lower operating costs and create less environmental pollution, but the fuel used by generators is at risk. If nuclear fuel is leaked, the surrounding environment will be devastated. At the same time, as the capacity of a nuclear power plant is huge, the impact on the power grid system will be great if the unit shuts down for maintenance. Building large scale hydropower station or pumped storage power stations could help meet the variation of consumption, thus enhance nuclear power plants' fuel efficiency. In this way, it would not only reduce the risks associated the nuclear fuel, but also cut the operating costs.

Comparées aux autres installations d'énergie renouvelables, les grands aménagements hydrauliques et les stations de pompage sont plus fiables et plus économiques, ont une plus longue durée de vie et jouent un rôle important pour le réseau électrique de puissance, grâce à leurs fortes capacités de stockage de l'énergie. Ainsi les fermes éoliennes, les centrales photovoltaïques et les centrales nucléaires pourraient bénéficier du support apporté par les grands aménagements hydroélectriques et les stations de pompage. Pour les aménagements nucléaires, cela réduirait les coûts d'exploitation et augmenterait la durée de vie des aménagements. Pour les fermes éoliennes et les centrales photovoltaïques, cela réduirait l'influence sur le réseau électrique et améliorerait la stabilité du système électrique.

Aujourd'hui beaucoup de pays à travers le monde développent les énergies renouvelables pour réduire le recours aux énergies fossiles, ceci pour respecter leurs obligations vis-à-vis des traités internationaux comme le Protocole de Kyoto. Les grands aménagements hydroélectriques avec de grandes retenues et les stations de pompage joueront un rôle de plus en plus important dans la stabilité du réseau électrique en se coordonnant avec les autres formes d'énergies renouvelables.

Par exemple en Europe, qui est à l'avant-garde du développement des énergies renouvelables, les pays sont interconnectés à travers le réseau électrique et l'équilibre entre offre et demande d'électricité se fait sur la base de pays à pays. Un pays avec de grandes quantités d'énergie hydraulique ajuste l'offre et la demande d'un autre pays, qui produit essentiellement à partir d'énergie éolienne et photovoltaïque. Par exemple la Norvège, où la production hydraulique représente 95% du total de l'énergie produite, soutient l'équilibre offre-demande au Danemark, où 20% de l'énergie produite est d'origine éolienne. En adoptant les sources en énergie renouvelable, dont la production est variable, il est nécessaire d'avoir d'autres formes d'énergie qui puissent absorber ces variations en modulant la puissance produite. Les régulateurs européens du réseau électrique rédigent de nouvelles règles pour traiter ce problème.

Au Japon, l'hydroélectricité représente environ 20% de l'énergie totale produite, bien qu'il y ait de nombreuses rivières. Il est ainsi important de faire appel de manière efficace aux capacités de régulation des centrales thermiques et des stations de pompage.

3.5.   LE MARCHÉ DE L'ÉLECTRICITÉ

En général les grandes installations hydroélectriques procurent des bénéfices significatifs aux systèmes électriques. Premièrement l'énergie potentielle de l'eau peut être stockée en grande quantité dans la retenue pendant les périodes de faible demande en électricité et être disponible lorsque la demande augmente. Ainsi la production hydroélectrique aux heures de pointe de consommation peut compenser les autres formes d'énergie moins flexibles comme les centrales thermiques, nucléaire ou à combustible fossile. Deuxièmement les faibles coûts variables de la production hydraulique, ainsi que sa capacité à assurer rapidement la charge, sont deux avantages par rapport aux groupes thermiques pour répondre aux fortes fluctuations du réseau électrique. Cette flexibilité est un avantage remarquable qui ne peut être égalé par les autres formes de production d'énergie. Ainsi les aménagements hydroélectriques sont les plus efficaces pour fournir des services au système électrique du fait de leur bonne flexibilité dynamique. Ainsi ils peuvent rapporter de larges bénéfices aux producteurs, dans la mesure où les services systèmes sont achetés et vendus sur un marché compétitif. De plus ces avantages permettent aux aménagements hydroélectriques de compléter les énergies renouvelables décrites au-dessus. À part ces aspects électriques, les aménagements hydroélectriques contribuent à beaucoup d'autres activités non commerciales ou sociales, et permettent de faire un usage complet de la 'valeur de l'eau'.

La marchandisation du secteur de l'énergie est une tendance générale à travers le monde. Depuis 1980, il a été introduit dans les pays développés et certaines zones entrainant la dérégulation et l'introduction de la concurrence, ceci afin d'améliorer l'efficience de l'industrie électrique. Aujourd'hui plusieurs pays ou zones ont établi leur propre marché de l'énergie, particulièrement en Europe : Grande Bretagne, Pays nordiques, France, l'Espagne et le Portugal etc. où existent des réseaux électriques. Également au Japon, aux USA : PJM (Pennsylvanie, Jersey et Maryland), Californie, Ontario, en Australie, en Nouvelle Zélande, etc. La Chine et d'autres pays émergeant explorent les voies possibles pour réformer leur marché de l'énergie.

Compared to the other renewable energy power plants, large hydropower stations and pumped storage power plants are the most reliable and economical, with the longest life cycles, and play an important role in power grids due to their high-capacity energy storage devices. Therefore, wind turbines, photovoltaic and nuclear power plants will benefit from the building of supporting large hydropower stations and pumped storage power plants. For nuclear power generators, it will reduce the cost of operating and extend the life of unit. For wind turbines and photovoltaic, it will not only reduce the impact on the power grid system, but also improve the stability of the system.

Nowadays many countries across the world are devoted to developing sustainable energy and reduce the usage of fossil fuels to comply with the regulations of treaties such as the Kyoto Protocol. Hydropower stations with big reservoirs and pumped storage power plants will play an ever more important role in power grids by coordinating with other sustainable energy.

For example, in Europe, which is at the vanguard of employing renewable energy sources, countries are connected to one another via power transmission lines, and the supply and demand are adjusted on a country-to-country basis. A country with large quantities of hydropower adjusts supply and demand in another country where large quantities of solar power and wind power are used for power generation. Norway, for example, where hydropower generation accounts for approximately 95% of total power generation, plays a role in adjusting supply and demand in Denmark where approximately 20% of total power is generated by wind power. Adopting renewable energy power sources needs to solve the problem of their varying power outputs. European regulators are writing new legislation to cope with this new issue.

In Japan, hydropower plants generate only approximately 20% of total power, even though there are numerous rivers. It is therefore important to make efficient use of the adjustment capacity of thermal power generation plants or pumped storage power plants.

3.5.   POWER MARKET

Generally speaking, large hydropower dams with reservoirs provide significant benefits to an electricity system. Firstly, the potential energy can be stored in large quantities within the reservoir during periods of low demand and be available when the demand rises. Thus, the production from hydropower at peak hours could compensate for other power sources with less flexibility, such as nuclear and thermal power plants. Secondly, the low variable costs of hydropower production and its short ramp-up time mean it is more rapid and less expensive than thermal generators when responding to drastic load fluctuations in the power grid. This flexibility gives it a remarkable advantage with which other generating sources cannot compete. Thus, hydro facilities are the most efficient sources of ancillary services because of their good dynamic flexibility and they may earn a substantial profit if ancillary services are purchased in a competitive market. These advantages also make hydropower a useful complement to the higher penetration of intermittent renewable energy sources described above. Apart from these aspects, hydropower is also used for many other non-commercial and social activities and makes full use of the "value of water".

Power sector marketization is the general trend for the electricity industry across the world. From the 1980s, it has been introduced into developed countries and areas through deregulation and the introduction of competition in order to improve efficiency in the electricity industry. At present, many countries and areas have established their typical electricity markets, especially in Europe: Britain, Nordic countries, France, Spain, Portugal, etc. where an electric net is settled. Also, in Japan, in the US: PJM (Pennsylvania, Jersey, Maryland), California, Ontario, in Australia and New Zealand etc. China and other emerging economies are also actively exploring route of power sector marketization reformation.

La méthode couramment utilisée pour la création d'un marché compétitif de l'énergie consiste à introduire la compétition sur les marchés de la production et de la vente de l'électricité, alors que le transport et la distribution restent régulés. La compétition à la production implique privatisation et restructuration du secteur des entreprises historiques, organisées verticalement pour créer deux ou plusieurs producteurs dans un marché de gros ouvert, marché également ouvert pour les nouveaux producteurs. Le marché de gros de l'électricité inclus un marché des contrats bilatéraux, un marché journalier, un marché infra journalier, un marché des services systèmes, un marché à terme, etc. Le centre où se déroulent les transactions entre les compétiteurs du marché de gros de l'électricité, est le cœur du marché de gros de l'électricité.

The current competitive power market method is based on introducing competition into power production and retail, while maintaining regulation for transmission and distribution. Competition in the field of power production means privatization and restructuring in the power production parts of original vertical integrated power companies and formation of two or more power producers on an open wholesale power market which is also open to new producers. The wholesale market includes a bilateral contract market, daily market, intraday market, ancillary services market and forward market, etc. The transaction centre where the competitive electricity wholesale transactions occur is the soul of the market.

4. EXPLOITATION OPTIMISÉE DES RÉSERVOIRS ET CENTRALES HYDROÉLECTRIQUES

Le développement de sociétés modernes augmente les besoins en énergie. Avec cette demande de plus en plus importante, la population est de plus en plus attentive à l'utilisation efficace de cette énergie. L'exploitation coordonnée des centrales hydroélectriques et des réservoirs peut améliorer significativement l'efficacité de l'utilisation de l'eau et optimiser la valeur de cette eau. L'exploitation intégrée de cascades d'aménagements hydrauliques est devenu une tendance mondiale.

4.1. CONTEXTE ET ÉVALUATION D'UNE EXPLOITATION INTÉGRÉE

4.1.1. Mise en œuvre d'une appréciation multicritères exhaustive

Avec le développement économique et social, l'utilisation de la ressource en eau passe de simple projet de conservation sur un site spécifique, à une utilisation de la ressource en eau exhaustive sur un bassin versant. Les réservoirs ont désormais une plus grande taille et doivent répondre à l'ensemble des fonctionnalités attendues du bassin versant. Toutefois à cause du développement économique et de l'augmentation de la population, des conflits entre la demande et l'offre de la ressource en eau se sont constamment intensifiés. Ceci est bien sûr lié à l'augmentation des demandes en eau et en énergie.

Les réservoirs en cascade sont en principe utilisés pour des objectifs multiples, comprenant la génération d'énergie, la gestion du passage des crues, l'approvisionnement en eau, la navigation, la gestion des sédiments et la protection écologique. Ces objectifs sont parfois en conflit les uns avec les autres, et parfois incompatibles (produire de l'énergie en été et irriguer les terres agricoles au même moment). L'utilisation et la régulation des retenues impliquent une maximisation des apports en eau et des bénéfices de la production d'énergie, tout en minimisant les risques liés au passage des crues et en satisfaisant les contraintes de protection environnementales. Ainsi il s'agit d'un problème typique d'optimisation multicritères, qui nécessite des décisions appropriées.

Ce problème d'optimisation multicritère est un processus de décision très complexe, qui implique de multiples aspects. On fait généralement l'hypothèse qu'il peut être décomposé en diverses caractéristiques additives dépendantes : la gestion des crues, la production d'énergie, l'approvisionnement en eau, la navigation, etc.

La valeur d'une gestion intégrée d'une cascade de réservoir peut ainsi être représentée par une fonction additive. La valeur de la gestion intégrée peut être décomposée et exprimée à travers différentes caractéristiques cibles. La fonction de valeur ou d'utilité définie à partir de cibles différentes, est destinée à mettre en œuvre la théorie de l'utilité dans un environnement défini ; on l'appelle également la fonction de valeur. La fonction de valeur utilise généralement les cibles définies pour la régulation du réservoir et les uniformise, jusqu'à ce qu'elles puissent être comparées via des valeurs physiques. Pour prendre en compte l'importance relative des cibles, des poids sont attribués selon un consensus à définir par les différents acteurs ou les autorités de contrôle. Les poids des différentes cibles sont ainsi plus objectives et proches de l'objectif de maximiser les bénéfices de l'utilisation de l'eau.

Cet outil est destiné à aider les responsables à prendre les bonnes décisions. Les personnes en charge ou responsable du processus, prendront une ou plusieurs décisions pour résoudre les problèmes à différentes échéances de temps. Bien qu'il y ait des différences liées à des méthodes propres à chaque pays, elles ont toutes le même objectif : optimiser le partage de l'eau et satisfaire ses différents usages.

4. INTEGRATED OPERATION OF HYDROPOWER STATIONS AND RESERVOIRS

The development of a modern society needs the support of power. As energy issues become more and more important, people become more concerned with energy utilization efficiency. The integrated operation of hydropower stations and reservoirs can improve the efficiency of water use significantly and realize the "value of water" to a large extent. Integrated operation has become a worldwide trend.

4.1. ANALYSIS AND ASSESSMENTS FOR INTEGRATED OPERATION

4.1.1. Multi-target comprehensive utilization and regulation

With social and economic development, utilization of water resources at a single water conservancy project in a specific location is gradually evolving into comprehensive utilization of the water resources of whole drainage basin. Reservoirs tend to have a larger scale and a full range of functions to respond to all of the expected needs. However, because of regional economic development and population expansion, contradictories between supply and demand of water resources have continuously been intensified. This is of course related to the increasing demands placed on water and energy.

Cascade reservoirs are normally used for diversified purposes, including power generation, flood management, water supply, navigation, sediment blocking and ecological protection. These purposes sometimes conflict with one other or are even incompatible (for example production of energy in summer and irrigation at the same time). Comprehensive utilization and regulation of reservoirs means maximizing water supply and power generation benefits, minimizing flood risks and meeting various environment protection requirements. Therefore, it is a typical multi-target optimisation issue that needs appropriate decisions.

This issue involves a very complex multi-target decision-making process, which touches on multiple aspects. It is generally assumed that it can be divided into the several dependent parts, such as flood management, power generation, water supply and navigation etc.

The value of integrated operation of cascade reservoirs can be described as an additive value function. The value function can be broken down and expressed as the value of various target properties. It generally utilizes the targets of reservoir regulation and uniformizes them until they can be compared via physical values. To take into account the relative importance of targets, weights are allocated according to the consensus made by stakeholders or legislative verdict. Weights of various goals determined in this manner are more objective and closer to the goal of maximizing benefits from general utilization.

This tool aims to help decision makers to take the right decisions. These people who are responsible for the process will take one or several decisions to solve the issue at different periods of time. Although there are differences, due to the methods adopted in each country, the goal is the same: optimizing distribution of water and fulfilling different usages.

4.1.2. Évaluation générale de l'exploitation d'aménagements en cascade

Avec la gestion multi-usage des aménagements hydroélectriques et des réservoirs sur un bassin versant, une attention considérable a été portée à la maximisation des bénéfices apportés par cette exploitation. Ainsi l'importance de l'évaluation de ces bénéfices est mise en lumière. Leur évaluation est nécessaire pour comprendre l'exploitation et la gestion des centrales hydroélectriques en cascade. Des modifications raisonnables des modes opératoires des diverses retenues sont nécessaires pour améliorer le bénéfice global de l'exploitation des aménagements.

(1) Respect du ratio de production en puissance

Le respect du ratio de production en puissance se réfère à la production en puissance réelle rapportée à la puissance maximale théorique de l'usine hydroélectrique. La puissance théorique correspond à la puissance maximale théorique de l'usine en prenant en compte certaines conditions aux limites. La génération d'énergie maximale théorique pour une année donnée, se réfère à la production maximale d'énergie basée sur le débit entrant dans la machine et dans le système d'optimisation de la gestion intégrée des aménagements. La production d'énergie maximum théorique est différente pour différentes conditions aux limites.

(2) Taux d'énergie supplémentaire de par la gestion des crues et l'optimisation d'exploitation des réservoirs

Le taux d'énergie électrique supplémentaire en provenance de la gestion du passage des crues et de l'optimisation du processus se réfère au ratio entre l'énergie électrique supplémentaire qui serait générée par un apport d'eau supplémentaire, par rapport à l'énergie électrique réellement produite. L'eau supplémentaire en provenance du passage des crues et de l'optimisation du processus est le volume d'eau supplémentaire résultant de la retenue de la crue, de l'optimisation en dynamique des contraintes de niveau pour le passage de la crue et des autres mesures de régulation. Il représente les bénéfices de la production d'énergie supplémentaire par une bonne gestion du passage de la crue.

(3) Taux d'amélioration de l'utilisation de l'eau par les centrales hydroélectriques

L'amélioration du taux d'utilisation de l'eau par les centrales hydroélectriques est un index dynamique très important. Il compare l'amélioration (ou la détérioration) de l'énergie produite réelle par rapport à celle qui était prévue.

La méthode est :

- Premièrement, on calcule la différence entre la production réelle et attendue, estimée à partir des débits entrants et du programme de production.

- Deuxièmement, on obtient le résultat par le ratio entre la différence de production et la production théorique.

Le ratio représente l'utilisation de la ressource en eau. Les opérateurs exploitent le réservoir ou l'aménagement hydroélectrique pour obtenir de meilleurs ratios, par exemple en atteignant de plus fortes charges hydrauliques, en utilisant de manière répétée les capacités de stockage, etc.

(4) Évaluation du management des crues

Les bénéfices du management des crues sont essentiellement sociaux. Ils consistent à éviter les déplacements de personnes, les pertes de propriété, de prévenir les arrêts d'activité industrielles et commerciales, etc. Les bénéfices sont difficiles à quantifier, mais il existe des méthodes qui permettent de traiter les conséquences d'une crue.

4.1.2. General assessments of cascade operation

With the multi-target integrated operation of cascade reservoirs and hydropower stations in the drainage basin, considerable attention has been placed on the maximization of general benefits linked to this operation. Thus, the importance of assessment for such general benefits is highlighted. These assessments are necessary for understanding a hydropower station's efficiency of regulation, operation and management. Reasonable modification of the operating modes of various reservoirs is necessary to increase the integrated operating benefits of the cascade.

(1) Fulfilment rate of power generation

The fulfilment rate of power generation refers to the ratio of actual power generation to theoretical maximum power generation of a power plant. Theoretical maximum power generation refers to a power plant's theoretical maximum power generation with certain boundary conditions. The theoretical maximum power generation of a specific year refers to the maximum power generation based on the actual water flows of the year through optimal regulation and tapping various factors to the utmost. Theoretical maximum power generation varies at different boundary conditions.

(2) Rate of additional electricity from flood regulation and optimization

Rate of additional electricity from flood regulation and optimisation refers to the ratio of additional electricity generated from additional water to the electricity actually generated. Additional water from the flood regulation and optimization is the extra water volume obtained by flood detention, dynamic control of flood water level limits and other regulation measures. It is expected to reflect the benefits of additional electricity gained from optimised management of floods.

(3) Water utilization improvement rate of hydropower stations

The water utilization improvement rate of hydropower stations is a very important dynamic index. It refers to the improvement ratio of actual power production to expected power production.

The method is:

- Firstly, calculate the difference between actual power production and expected power production, which is calculated by model according to actual inflow and operation diagram.

- Secondly, get the result of the ratio between the difference and expected power production.

The ratio reflects the better utilization of water resources. The producers run reservoirs and hydropower stations in an optimised way to achieve better rates, such as gaining higher water head, repetitively utilizing limited storage capacity, etc.

(4) Assessment of flood management

Flood management benefits are mainly social ones, which include avoiding relocation of residents and property losses, preventing interruptions in industrial and commercial activities, etc. They are hard to quantify, but there are methods to assess impact of flood.

(5) Évaluation des économies d'énergie et de la réduction des gaz à effet de serre

Cette évaluation décrit l'amélioration de la production hydroélectrique à travers l'optimisation de la production des centrales enchaînées. Les bénéfices incluent les gains par rapport à une production qui sinon se ferait à partir de combustibles fossiles (dont le charbon, une grande source d'énergie dans le monde), et également par la réduction des émissions de CO_2, NO_x et SO_2.

Ces évaluations peuvent aider à clairement identifier les bénéfices sociaux et environnementaux apportés par une exploitation intégrée, ainsi que de mettre en évidence les manques en cours d'exploitation, qui permettent d'amélioration le processus.

4.2. OUTILS INFORMATIQUES POUR LE CONTRÔLE DES AMÉNAGEMENTS HYDRAULIQUES ET DES RÉSERVOIRS

Avec le développement des techniques de l'information, les outils techniques qui permettent de contrôler et d'exploiter les aménagements hydroélectriques sont de plus en plus sophistiqués et fiables.

Les systèmes de surveillance, de contrôle et d'acquisition de données (SCADA) et les systèmes de gestion des retenues (SGR), sont devenus des outils communs pour le contrôle et l'exploitation des centrales hydroélectriques et des réservoirs. Lorsque des données ont été acquises à travers ces outils, le contrôle de leur validité est nécessaire (critique), avant de les introduire comme données d'entrée dans les outils SCADA et SGR ou de les communiquer au public. Dans certains pays le module SGR est intégré dans le SCADA.

4.2.1. Système de contrôle informatisé des centrales hydrauliques (SCADA)

4.2.1.1. Évolution du SCADA (Chine)

SCADA fournit des données en temps réel automatiquement des centrales électriques et alimente le système d'exploitation du système intégré (EMS : Electrical Management System). Le SCADA surveille, contrôle et analyse le processus de production d'énergie d'une centrale hydroélectrique. Il fournit les informations de production. Avec le développement des technologies de communication entre ordinateurs et des équipements de contrôle industriels, le SCADA est mis en œuvre en trois étapes :

- Première étape, phase expérimentale : quelques centrales hydroélectriques réalisent des mesures automatiques et effectuent le traitement de données pour le processus de production.

- Deuxième étape, phase opérationnelle : Le SCADA est impliqué dans des boucles de contrôle opérationnelles.

- Troisième étape, phase de maturité : les SCADA continuent de se développer et se popularisent dans le monde entier.

Aujourd'hui, ce sont maintenant des systèmes complets, intégrés dans l'organisation qui permettent non seulement de suivre le processus industriel, mais également de réaliser une surveillance des ouvrages et des matériels, d'en déduire les plans de maintenance et ainsi éviter tout arrêt ou rupture des ouvrages ou des équipements. Dans ce bulletin le SCADA est focalisé principalement sur la régulation des réservoirs et des aménagements hydroélectriques. Différents pays ont développé des SCADA selon leur propre méthode ; des généralités sont présentées ci-après.

(5) Assessment of energy saving and emission reduction benefits

This assessment describes the benefit of increased power hydroelectric production from optimization. The benefit is composed of energy saved which could be evaluated by standard coal, as well as emissions reduction of CO_2, NO_x and SO_2.

These assessments can help us clearly evaluate commercial and social benefits brought by integrated operation and find deficiencies during operations so that more improvements could be made.

4.2. INFORMATION TECHNOLOGY TOOLS FOR MANAGING HPPS AND RESERVOIRS

With the development of information technology, the information technology tools used for managing and operating HPPs and reservoirs are more and more advanced and reliable.

Supervisory Control and Data Acquisition System (SCADA) and Reservoir Dispatching System (RDS) are common information technology tools for managing and operating HPPs and Reservoirs. When data have been collected through these tools, checking their validities is critical before inputting them into function modules or presenting them to the public. In certain countries, RDS is integrated into the SCADA.

4.2.1. Supervisory Control and Data Acquisition System (SCADA) of power generation

4.2.1.1. Evolution of SCADA

SCADA is a real-time data source for the automation of electricity power systems and provides a large amount of real-time data for Energy Management Systems (EMS). The SCADA works to monitor, control and analyse the power generation process of a hydropower station and provides production information. With the development of computer communications technology and industrial control technology, SCADA basically develops in three stages:

- First stage – trial stage: some hydropower stations have realized automatic measurement and data processing in production. The SCADA starts to be used to record and monitor the operating parameters of the stations.

- Second stage – operational stage: the SCADA involved in closed-loop control.

- Third stage – maturing stage: the SCADA continues to develop as it gradually becomes popularized across the world.

Nowadays, the system is not only surveying the operating process, but is also aimed at surveying behaviour of works and equipment to detect any abnormal situations and produce maintenance programs to prevent breakdown or failure of equipment. In this bulletin, the SCADA is mainly focussed on integrated operation of hydropower stations and reservoirs. Different countries usually have their own characteristics in the process of establishing SCADA, and here just introduce the general situation.

4.2.1.2. Introduction aux fonctions générales du SCADA

Un système SCADA capte en temps réel des informations à partir du système d'information de la conduite, mais aussi des instructions de régulation à travers une interface de communication. Il génère des séquences d'évènements (SE) et des alarmes à travers diverses méthodes d'analyse des données. Il les affiche à l'opérateur, sous la forme de graphes, de courbes et d'enregistrements sonores. En même temps il conserve les données historiques afin de pouvoir réaliser des statistiques, de tracer les historiques d'accidents et d'analyser ultérieurement un dysfonctionnement. Des personnes autorisées peuvent surveiller, contrôler et ajuster la production en temps réel à travers une interface Homme-Machine (IHM).

De plus le SCADA dispose de systèmes de contrôle automatiques de la puissance (AGC), de contrôle automatique de la tension (AVC) et d'autres applications spécifiques, pour réduire la charge de travail des opérateurs et améliorer la qualité de l'électricité produite.

4.2.1.3. Structure du système de surveillance de centrales enchaînées sur un bassin versant

Les aménagements hydroélectriques sont en règle générale éloignés des centres de consommation. Ainsi il est nécessaire d'adapter les opérations fortuites pour réduire le nombre de personnel sur le site et donc augmenter la productivité de la centrale. Un système SCADA doit être développé pour surveiller et contrôler l'ensemble des aménagements hydroélectriques sur un même réseau hydraulique.

Les SCADA d'aménagements hydrauliques à l'intérieur d'un bassin versant a généralement trois couches successives: une couche groupes, une couche aménagement et une couche centre de supervision et de contrôle centralisé. Sur la couche groupe, le système est réparti en fonction des objets, un groupe générateur forme une unité de contrôle local. Sur les couches des aménagements et du centre de supervision centralisé, le système est décrit selon les fonctions; différentes fonctions sont déployées aux différents nœuds à l'intérieur de la grille du système, pour réaliser les tâches de manière collaborative.

Les différentes couches et les applications sont interconnectées à travers un bus informatique ou un réseau Ethernet. Pour une meilleure fiabilité et efficacité, la grille et le hardware ont une configuration redondante. Le protocole de communication standard international devrait être appliqué entre le SCADA aménagements intégré et les SCADA aménagements, de manière que les aménagements puissent être reliés au SCADA centralisé à moindre coût.

4.2.1.4. Développement du SCADA, tendances

L'hydroélectricité est une des sources d'énergie majeure pour le réseau électrique. Pour construire un réseau électrique intelligent (smart grid) et robuste, il est nécessaire d'améliorer encore la fiabilité, la qualité de l'électricité produite et le taux d'utilisation de l'énergie hydroélectrique. En conséquence, le SCADA doit satisfaire les standards sur des réseaux électriques intelligents. De plus il y aura des applications plus diversifiées et avancées du SCADA. En particulier la recherche sur l'exploitation économique, les bases de données sur l'historique des paramètres, l'analyse de l'état des aménagements hydroélectriques en cascade, pourront procurer un support technique à l'exploitation de ces aménagements.

4.2.1.2. Introduction to the general functions of the SCADA

The hydropower station SCADA realises real-time production information exchange and receiving of regulation instructions via a communication interface. It generates a sequence of events and alarm information through various data processing methods. It displays them to the operator via graphs, curves and audios. Meanwhile, it also saves historic data as the basis for statistics, accident tracing or posterior troubleshooting. Authorized operators can monitor, control and adjust the production process on a real-time basis via the human-machine interface.

Moreover, the SCADA normally has Automatic Generation Control (AGC), Automatic Voltage Control (AVC) and other advanced applications in order to reduce the workload of operators and increase the quality of electric power generation.

4.2.1.3. Structure of monitoring systems of cascade hydropower stations in drainage basins

Hydropower stations are normally located far away from consumption areas, therefore it is necessary to adapt the unattended operation in order to reduce the number of staff in the field and increase productivity. A cascade SCADA shall be developed to monitor and control the whole cascade hydropower station.

The SCADA of the cascade stations within drainage basins normally has three layers, namely a field layer, a station layer and a centralized control layer. On the field layer, the system is distributed according to the objects, one generator unit forms a Local Control Unit. On the station and centralized control layers, the system is distributed according to functions and different functions are deployed on different nodes within the network of the system to complete tasks collaboratively.

Various layers and applications are interconnected via field bus or Ethernet. For higher system reliability and efficiency, the network and hardware shall have a redundancy configuration. The international standard communication protocol should be applied for the communication between cascade SCADA and station SCADA so that various stations of the cascade can be connected to the cascade SCADA at lower cost.

4.2.1.4. SCADA development trends

Hydropower is one of the major energy sources for electric power grids. In order to build a smart and robust electric power grid, it is necessary to further increase the reliability, electricity quality and utilization ratio of hydropower. As a result, SCADA needs to satisfy the standards of smart power grid. Moreover, there will be more diversified advanced applications of SCADA. In particular, research on economical operation, historical data mining and status analysis of cascade hydropower stations will provide technical support for the operation of hydropower stations.

4.2.2. Système de gestion des réservoirs (SGR)

4.2.2.1. Définition d'un SGR

Le SGR est un important élément du système de management de l'énergie (EMS). Sa fonction principale est de fournir un support et des services à la régulation, à l'exploitation et à la gestion du réservoir. En particulier, il surveille, prévoit et régule l'exploitation de la retenue. En un mot, il offre un support technique pour garantir une exploitation sûre et efficace économiquement de la ou des retenues.

À partir d'acquisition de données avancée, de technologies de communication et d'applications, le système peut acquérir précisément les données hydrauliques, météorologiques, d'état du réservoir et d'état de la production électrique. Il utilise des ressources d'ordinateur et des théories d'optimisation mathématiques pour des prévisions météorologiques en ligne, une régulation optimisée, des calculs hydrauliques et fournit des plans de régulation optimisés, qui couvrent le passage de la crue, le programme de production, la navigation, l'irrigation et l'approvisionnement en eau potable, pour réaliser les manœuvres automatiques et efficaces des aménagements hydroélectriques et du système électrique.

4.2.2.2. Structure du SGR

Le SGR a une structure redondante (réseaux duals). Les deux réseaux sont complètement indépendants et dotés des mêmes applications, pour s'assurer qu'aucun défaut en un point du réseau, ne puisse mettre en défaut la fiabilité opérationnelle du système. Afin d'améliorer la fiabilité et la sûreté du système, une détection d'intrusion et un pare feu devrait être implémenté dans le réseau. Globalement les questions de sécurité des données et des applications informatiques sont de plus en plus importantes pour les aménagements hydroélectriques.

Le SGR comprend quatre parties, le module applicatif de la régulation réservoir, l'interface des données externes, les mesures télé-opérées, la communication des données. La structure est décrite à la Figure 4.1.

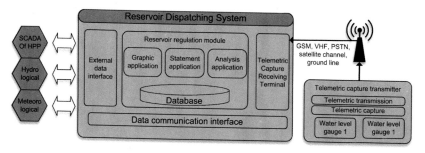

Figure 4.1
La structure du SGR

Le module applicatif de la régulation réservoir, permet la régulation et l'exploitation du réservoir et fournit des services météo hydrologiques, des données de mesures électriques, de navigation, d'énergie produite et d'autres services sur la régulation des aménagements enchaînés.

4.2.2. Reservoir Dispatching System (RDS)

4.2.2.1. Features of RDS

The RDS is an important constituent of the Energy Management System. Its core function is to provide support and services for the regulation, operation and management of the reservoir, especially at aspects such as monitoring, forecasting and regulating. In a word, it offers technical supports to ensure safe and cost-efficient operations of the reservoirs.

The system can accurately acquire the hydrological, meteorological, reservoir and generation information through advanced data capture, communication and computer application technologies. It uses computer technology and mathematic optimization theories for online hydrological forecasts, optimized regulation, and hydraulic calculation and provides optimized regulation plans that cover flood control, power generation, navigation, irrigation and water supply to realize the automatic and efficient regulation of hydropower stations and power grids.

4.2.2.2. Structure of RDS

RDS has a redundant dual-network system. The two networks are completely independent and furnished with corresponding applications to ensure that single-point fault will not affect the operational reliability of the system. In order to increase reliability and security, an intrusion detection system and a firewall should be deployed in the network. Globally, cyber and information security is becoming more and more important to hydropower stations.

RDS comprises four parts, i.e. reservoir regulation application module, external data interface, telemetric capture receiving terminal and data communication. Its structure is as follows:

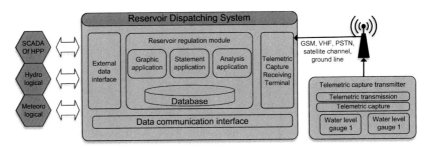

Fig. 4.1
Structure of RDS

The reservoir regulation application module enables the regulation and operation of the reservoir and provides hydro-meteorological services, electric data communication, navigation, power generation and other cascade reservoir regulation services.

Le module applicatif de la régulation réservoir comprend une interface Homme-Machine, une couche processus industriel et une couche de support post-traitement. L'interface Homme-Machine affiche graphiquement les informations concernant l'hydrologie, la météorologie, la puissance générée, les groupes de production et les vannes de commande. Sur le même écran sont affichés l'état des organes, les statistiques, les analyses et les résultats de calcul. La couche processus industriel complète principalement le processus de gestion du réservoir. La couche de support post-traitement résume et intègre les données acquises par le sous-système mesures télémétriques. Ces systèmes doivent être contrôlés et sont alimentés par le module de communication des données, ils sont stockés dans des fichiers et l'exploitation du système est surveillée.

L'interface pour les échanges des données permet des échanges avec le système de régulation du réseau électrique, le système hydrologique, le système météorologique, le système d'exploitation centralisé, l'administration du bassin versant et l'administration du réseau électrique.

La collecte des mesures télémétriques se fait sur le terminal de réception : les données pluviométriques, les niveaux d'eau, les taux de saturation des sols, les vitesses et directions du vent, ainsi que d'autres données hydrométéorologiques sont collectées. Elles sont retransmises par le système télémétrique, qui les sauvegarde dans la base de données de régulation du réservoir et sur la plateforme opérationnelle.

Le système de communication des données est capable d'échanger des données avec d'autres systèmes de régulation de réservoir ou d'autres tierces parties au niveau de la base de données.

4.2.2.3. Le sous-système de mesure télémétrique du régime des eaux

La télémétrie du régime des eaux est une partie inséparable du SGR et ces données une ressource pour l'exploitation coordonnée des retenues, de l'écoulement des eaux et de la production électrique. Il est très important d'établir un réseau télémétrique précis, fiable et en temps réel du régime des eaux.

La détection automatique du régime des eaux et la technologie de traitement des données est une combinaison de savoir-faire multidisciplinaires comprenant le capteur, sa surveillance et son contrôle, la communication entre le capteur et l'application informatique qui traite les données pour les prévisions hydrologiques. Les appareils télémesurés enregistrent les précipitations, les hauteurs d'eau, l'évaporation, l'humidité du sol, la température, l'humidité de l'air, la vitesse du vent et sa direction, ainsi que d'autres paramètres d'hydrologie, de météorologie et de qualité des eaux. A partir de ces données, on obtient en temps réel, le régime des eaux, la pluviométrie et des données météorologiques pour l'exploitation du réservoir et les prévisions de crues.

Après le recueil des données, la station télémesurée les transmet à travers plusieurs canaux de communication. La station centrale (une agrégation d'hardwares et de softwares rassemblés au centre de dispatching et de contrôle) reçoit les données de toutes les stations télémesurées à travers le software de collecte des données, les analyses selon un protocole défini au départ, pour finalement les transcrire dans le système d'exploitation du réservoir.

Le système télémesuré des Trois Gorges en Chine a fait l'objet de plusieurs modifications depuis 2003, date à laquelle il a été mis en place : expansion du réseau, mise à jour des logiciels, améliorations. Jusqu'à présent, 633 stations télémesurées ont été installées, elles couvrent un bassin versant de 580 000 km^2.

The reservoir regulation application module comprises a human-machine interface, a business logic processing layer and a background support layer. The human-machine interface displays hydrological, meteorological, power generation, generator units and sluice gate information graphically together with the necessary status display, statistics, analysis and calculations. The business logic processing layer mainly completes the reservoir regulation process. The background support layer summarizes and integrates the data captured by the telemetric subsystem. These systems have to be controlled and provided by the data communication module, they are kept on file and system operation monitored.

The external data interface allows data exchanges between the electricity regulation system, hydrological system, meteorological system, office system, drainage basin administration and power grid administration.

The telemetric collecting & receiving terminal receives the rainfall, water level, soil moisture content, wind speed, wind direction and other hydro-meteorological data transmitted back by the telemetric collecting device and saves them in the database of the reservoir regulation and application platform.

Data communication interface is liable for the data exchange with other reservoir regulation systems and other third-party systems on the database level.

4.2.2.3. Water regime telemetric subsystem

Water regime telemetry is an inseparable part of RDS and the data support to the integrated operation and control system of water and electricity. It is very important to establish an accurate, real-time and reliable water regime telemetry subsystem.

Water regime automatic detection and report technology is a combination of multidisciplinary applications, including sensor, survey and control, communication, computer application and hydrological forecast. The telemetric capture device acquires the rainfall, water level, evaporation, soil moisture content, temperature, humidity, wind speed, wind direction and other hydrological, meteorological and water quality data, hence offers real-time water regime, rain condition and meteorological data for reservoir regulation and flood forecasting.

After data collection, the telemetric station transmits the collected data via multiple communication channels. The central station (a collective of hardware and software deployed in the dispatch and control center) receives the messages from the telemetric station via the data collecting software, interprets the messages according to the predefined message protocol and finally writes the data into the reservoir regulation system database.

The TGP water regime telemetric system has undergone expansion, upgrading and improvement a couple of times since it was launched in 2003. So far, 633 telemetric stations that cover a total area of 580,000km^2 have been established.

4.2.2.4. Tendance pour le développement du SGR

Le développement du SGR est guidé par les fortes demandes du marché, l'évolution constante des technologies et les recherches approfondies. Le développement comprend en quatre étapes :

La première étape implique la réception des données, leur traitement et leur stockage. La seconde implique principalement les applications du système de configuration. La troisième étape implique les applications du middleware. La quatrième concerne les applications profondes ou cœur du système. Elles concernent un support plus profond comprenant les prévisions, les applications d'exploitation de la retenue, les autres régulations de la retenue, ainsi que la capacité à diversifier les représentations graphiques, notamment en utilisant les ressources d'un système d'informations géographiques (SIG).

Le futur SGR sera développé en prenant en compte complètement les prévisions quantitatives de précipitations, les prévisions de crue, les méthodes et théorie de prise de décision pour l'exploitation des réservoirs et aménagements, l'analyse de risque pour les prises de décision d'exploitation. Les technologies de systèmes d'information géographiques (SIG), les technologies d'optimisation de l'exploitation, les algorithmes de gestion des crues sur des aménagements en cascade seront intégrées dans les évolutions, de même que l'exploitation des aménagements, la gestion du réseau électrique, la protection de l'environnement et autres bénéfices liés à l'exploitation. Ces développements seront graduellement appliqués à la gestion des aménagements en cascade et deviendra une partie indissociable d'un aménagement hydroélectrique performant.

4.2.2.4. RDS developments trends

The development of RDS is driven by strong market demands, ever-developing technologies and in-depth research. Its development basically undergoes four stages.

The first stage mainly involves data reception, processing and storage. The second stage mainly involves application of a system configuration frame. The third stage mainly involves the application of middleware. The fourth stage is the in-depth application stage which is mainly reflected by the platform's deeper and more complete supports for forecast, regulation and other reservoir regulation applications as well as the availability for more diversified display methods including a Geographic Information System (GIS).

Future RDS shall be designed for giving full consideration to quantitative rainfall forecasts, flood forecasts, smart regulation decision-making theories and methods, regulation decision-making risk analysis and GIS technology and with full reference to the optimized regulation technologies and algorithms of cascade reservoir flood control, power generation, electricity market, environment protection and other general benefits. It will be gradually applied in cascade reservoir regulation and become an inalienable part of a smart hydropower plant.

5. CONCLUSION

1. Le développement de l'hydroélectricité joue un rôle significatif et en augmentation dans l'industrie de l'énergie. Dans le contexte actuel de crise de l'énergie et de problèmes écologiques environnementaux, l'utilisation de nouvelles formes d'énergie est devenue un point sensible dans le monde. L'hydroélectricité est une forme d'énergie de haute qualité, son exploitation à travers le monde, particulièrement dans les pays en développement, suscite un intérêt très répandu. Les aménagements en cascade, les réservoirs et les stations de pompage sont les principales formes d'exploitation hydraulique. Ils sont largement utilisés pour produire de l'énergie, irriguer, contrôler le passage des crues, permettre la navigation, assurer l'approvisionnement en eau industrielle, etc. Ils sont d'un grand secours pour le développement économique et social, mais également contribuent de manière très significative à la lutte contre les gaz à effet de serre et à la promotion du développement de l'énergie à basse émission de carbone dans le monde.

2. Les aménagements hydroélectriques et les stations de pompage ont une grande capacité d'ajustement. Comparés aux autres nouvelles formes d'énergie, les grands aménagements hydroélectriques et les stations de pompage sont les moyens de stockage de l'énergie les plus fiables et les plus économiques pour les réseaux électriques de transport de l'énergie. Leurs avantages sont, une production en puissance stable, un démarrage et une prise de la charge très rapide. Associées aux énergies éolienne, photovoltaïque et nucléaire, les centrales hydroélectriques peuvent compenser efficacement les déficiences de ces dernières et ainsi réduire les risques de rupture de charge sur les réseaux électriques.

3. L'exploitation intégrée des aménagements hydroélectriques et des réservoirs est impérative. L'exploitation intégrée des aménagements hydroélectriques et des réservoirs peut améliorer l'utilisation efficace de la ressource en eau, garantir la sûreté à l'aval lors du passage des crues, optimiser les coûts de production d'énergie et s'assurer que chaque centrale hydroélectrique fonctionne d'une manière appropriée et économique. Ils jouent un rôle très important dans la sûreté et l'exploitation stabilisée des réseaux de transport de l'électricité. L'exploitation intégrée des aménagements hydroélectriques et des réservoirs est devenue la tendance de l'exploitation hydraulique. Beaucoup de pays ont développé ou vont développer une plateforme d'exploitation intégrée SCADA (Système de contrôle et d'acquisition de données). Il en va de même des systèmes de gestion des retenues (SGR), des systèmes de télétransmission des données météorologiques et hydrologiques, des systèmes de communication, etc. Ceci pour améliorer la conduite des aménagements et finalement l'efficacité de la production.

Dans cette période de transformation de la production de l'énergie et de développement rapide des nouvelles formes d'énergie, l'hydroélectricité avec ses qualités uniques présente une grande opportunité de développement.

5.   CONCLUSION

1. Hydropower development plays an increasing significant role in energy industry. In the context of energy crisis and environmental ecological problems, the exploitation and utilization of new energy has become a hot spot in the world. Hydropower is a high-quality energy. Its exploitation and utilization across the world especially in the developing countries has got widespread concern. Cascade hydropower stations and reservoirs is a main form of hydropower exploitation, which are widely used in the aspects of power supply, irrigation, flood control, navigation, industrial water supply and etc. Hydropower not only provide a great help for economic and social development, but also are of great importance to alleviate the greenhouse effect and promote low-carbon development of the world.

2. Hydropower stations and pumped storage power plants have a great adjustment ability. Compared with other new energy, large hydropower stations and pumped storage power plants are the most reliable and economical energy storage devices in the power grids and have the advantages of stable power output, quick reaction and fast ramp-up. By coordination with wind energy, photovoltaic and nuclear power plants, hydropower stations and pumped storage power plants could make up for their deficiencies efficiently and reduce the risks of power grids.

3. Integrated operation of hydropower stations and reservoirs is imperative. As an efficiently technology measure, integrated operation of hydropower stations and reservoirs can improve the utilization efficiency of water resources, ensure the safety of downstream from flood risks, save production cost and make sure that every hydropower station operates in an appropriate and economical way. It has played a very important role in the safety and stable operation of power grids. Integrated operation of hydropower stations and reservoirs has become a trend of hydropower exploitation; many countries are devoted to establish the operation platform integrated with Supervisory Control and Data Acquisition System (SCADA), Reservoir Dispatching System (RDS), Water Regime Telemetric System, Communication System and etc. to improve the management level for higher production efficiency.

In the special period of transformation of global energy structure and the rapid development of new energy, hydropower will get attention with its unique advantages and get a new greater development opportunity.

6. ÉTUDES DE CAS

Les cas présentés ci-après ont été produit par dix pays et sont reliés aux aspects évoqués au-dessus, mais les points présentés sont centrés sur les caractéristiques différentes à chaque pays. Certains cas sont similaires et nombreux comme la prévision hydrologique, l'utilisation et la régulation globale des ouvrages pour des objectifs multiples, la surveillance centralisée et l'acquisition de données, etc. Cependant il y a également des différences. Certains cas se concentrent sur un ou deux aspects. Par exemple la contribution japonaise introduit principalement la production hydroélectrique, la contribution suisse se concentre sur le management du passage des crues, la contribution de la Corée est centrée sur le management de la ressource en eau pendant les étiages et la contribution brésilienne présente le contexte général et les enjeux qui doivent être relevés. Les détails de chaque étude de cas sont retranscrits ci-après par ordre alphabétique des pays.

Les études de cas ne sont pas traduites.

- La production hydroélectrique au Brésil

- La cascade des Trois Gorges-Gezhouba en Chine

- La cascade de l'Ariège en France

- La cascade de la rivière Dez en Iran

- La cascade de la rivière Kiso au Japon

- Les mesures d'adaptation contre les crues en Corée

- Réservoirs hydroélectriques en cascade au Nigeria

- La cascade hydroélectrique des rivières Volga et Kama en Russie

- La haute vallée du Rhône en Suisse

- L'aménagement de la vallée de la rivière Tennessee aux USA

6. SPECIFIC CASES

The following cases in ten countries are related to the contents described above, but their emphases are different according to their unique conditions. Some cases are similar and plentiful, referring to hydrological forecasting, multi-target comprehensive utilization and regulation, supervisory control and data acquisition etc. However, there are also differences. In some cases, the emphasis is on one or two aspects. For example, the case from Japan mainly introduces generation operations, the case from Switzerland explains flood management, the case from Korea introduces water resource management in drought and the case from Brazil introduces its basic information and challenges. Details of all cases are shown below, and the ten cases are ranged alphabetically.

6.1. HYDROPOWER GENERATION IN BRAZIL

6.1.1. Introduction

Brazil holds 12% of the planet's fresh water and has the third largest hydroelectric potential in the world (250,000MW). Today Brazil is the world's second largest producer of energy from hydroelectric plants and continues exploring its hydroelectric potential, but the socio-environmental policies and legislation began to restrict the construction of new projects, particularly the ones with large reservoirs.

The Brazilian Integrated Power System is primary in charge of the country's hydro and thermal power generation with 139 GW of total installed capacity. It covers 2/3 of the total country area of 8.5 Million square kilometres and supplies 97% of the total consumption of the country. The hydro part of the system distributed among 8 large hydrographical basins, the Amazon being the biggest one, consists of 69 multi annual or seasonal regulation reservoirs and 144 hydropower plants with an installed capacity of 86.7 GW by the end of 2014. The integrated system optimizes the operation of plants, sources and costs resulting in important synergetic benefits for the consumers.

6.1.2. Main socio-environmental restrictions

Brazil has modern legislation and a consolidated institutional framework for the regulation of projects and activities that may affect the environment, such as power generation projects. The main socio-environmental issues and restrictions affecting the development and implementation of a hydro projects in Brazil can be summarized as follows:

- The very restrictive socio environmental policies, laws and plans.

- The issues related to the resettlement of populations.

- The legal environmental protection areas.

- The legal Indian reservations.

- The national and international pressure applied by non-governmental organizations and other social groups.

- The social and economic effects during construction time.

- The social and economic effects of the demobilization of construction on the project region.

6.1.3. *The present situation of reservoirs and future consequences*

There is a rapid growth of energy demand however due to the strict socio-environmental restrictions. The Brazilian power sector authorities have been given preference to the run-of-river hydropower stations with daily or weekly storage capacity. The official generation expansion planning forecasts an increase of 45,000 MW of the hydro installed capacity from 2014 to 2023. However, this planning considers that the correspondent energy that can be stored in associated reservoirs is estimated to increase only by 7,000 MW.

As a consequence, Brazil is giving up some important social and economically benefits. The lack of large reservoirs increases the complexity of the integrated operation of the system. At this moment in the very dry years of 2013, 2014 and 2015, the Brazilian Integrated Power System is clearly being affected by the lack of additional storage energy capacity reservoirs, impacting its reliability, costs and risk of insufficient generation capacity.

However it is important to mention that the socio-environmental issues have always been considered in a very responsible way in the large hydroelectric projects that have been implemented in Brazil as well as there are countless examples of important development in the regions where large hydroelectric projects were built.

6.2. THREE GORGES-GEZHOUBA CASCADE COMPLEX – A CASE STUDY IN CHINA

6.2.1. *Introduction*

6.2.1.1. *Climatic and Geographic Conditions*

The Yangtze River is the longest river in China, with a total length of approximately 6,380 km and a total drop of 5,400m, covering an area of 1.8 million km². It discharges more than 960 billion m³ of water into the sea a year on average and accounts for 36% of China's total water resources.

In terms of terrain characteristics, the area of the upper reaches of the Yangtze River features a complex topography. In terms of climatic characteristics, due to the wide differences in terrain, topography and altitude within the watershed, there are two significantly different climate zones: a plateau monsoon climate zone and a subtropical monsoon climate zone. The bulk of the plateau monsoon climate zone is located in the upper and middle reaches of the Jinsha River and in the upper reaches of the Mintuo River. Within this climate zone, the annual precipitation ranges from 200mm to 600mm. The subtropical monsoon climate zone includes all the other watersheds and the annual precipitation range from 800mm to 1,600mm.

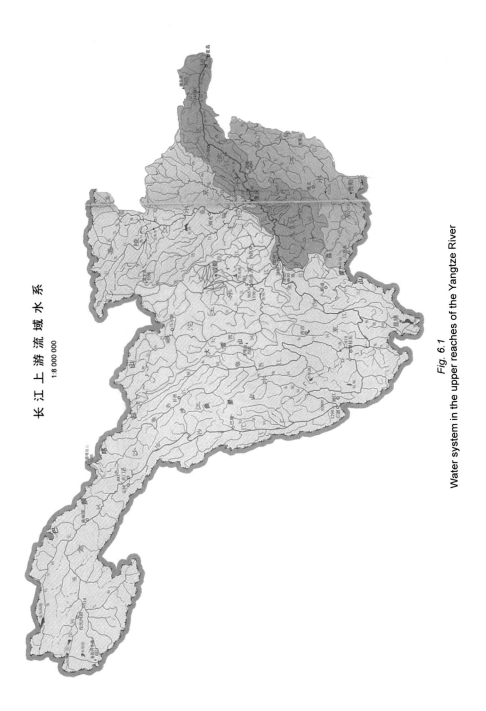

长 江 上 游 流 域 水 系

1:8 000 000

Fig. 6.1
Water system in the upper reaches of the Yangtze River

6.2.1.2. Runoff Characteristics

The Three Gorges-Gezhouba Cascade Hydropower Complex (TGP-GZB) is located in Yichang City, Hubei Province. The main stream of the Yangtze River has a large, steady annual runoff, with an average annual flow of 14,100m³/s and annual runoff of 446 billion m³ from 1878 and 2010. The trends of change remain steady over the long series.

The runoff in the watersheds upstream from TGP on the Yangtze River primarily comes from precipitation. The pattern of annual runoff distribution is similar to the pattern of precipitation, and distribution is uneven over the year. The flow of the trunk stream of the Yangtze River in the flood season accounts for 70% to 75% of the river's total annual flow, the flow of the trunk stream in the upper reaches is mainly concentrated in the period of June-September.

6.2.1.3. The TGP-GZB Cascade Complex

The TGP is a multi-objective development project designed to produce integrated benefits in terms of flood control, power generation and navigation. The reservoir has a normal storage level of 175m, a crest elevation of 185m, and a maximum height of 181m, with a total storage capacity of 39.3 billion m³ and a flood-control storage capacity of 22.15 billion m³. The reservoir can significantly improve the maneuverability and reliability of flood control operations in the middle and lower reaches of the Yangtze River. The Three Gorges Hydropower Plant has a total installed capacity of 22,500MW and is equipped with 32 mixed-flow (Francis) water-turbine generating units.

The GZB, located 38km downstream from the TGP, is the world's largest run-of-river hydropower station with a low water head, a large flow and a capacity of 2,910MW.

6.2.2. Comprehensive operation and dispatching

6.2.2.1. Precipitation forecast and hydrological forecast

(1) Precipitation Forecasting

The professional meteorological observatory mainly forecasts short-term rainfall zones, medium-term rainfall processes, and long-term rainfall trends in the upper reaches of the Yangtze River. At present, for rainstorms causing floods in the upper reaches of the Yangtze River, forecasts for 1–3 days are accurate and reliable; precipitation process forecasts for 4–7 days are useful; forecasts for 8–10 days may not be accurate, but can still be used as a guide for future precipitation trends.

(2) Short- and Medium-term Hydrological Forecasting

Hydrological forecasters provide 12h, 24h and 48h (3–7 days) forecasts of inflow to the Three Gorges Reservoir. The main rainfall runoff forecasting models for the Three Gorges watershed include the Xin'an River Model, the Water Tank Model, and the API Model. The main forecasting models for the flow on the Three Gorges section of the river include the Muskingum Flow Routing Method and the Hydrodynamics Model.

(3) Long-term Hydrological Forecasting

Long-term hydrological forecasting is intended to facilitate the preparation of long and medium-term power generation plans, reservoir drawing-down schemes, reservoir flood-season-coping schemes, and reservoir storage schemes. Such forecasting mainly includes monthly average reservoir inflow forecasting, flood season (April to October) forecasting, annual forecasting etc. The main long-term hydrological forecasting models currently used include the autoregressive model, the multiple regression model, and the threshold autoregressive model.

6.2.2.2. Operating methods and power transmission

These TGP generating units are located in different sections and are operated independently of each other without any electrical connection. The GZB is equipped with 21 generating units distributed in two independent powerhouses.

Electric power from TGP is consumed by ten provinces and municipalities in the economically vibrant central, eastern and southern parts of China. Electrical power from GZB is all consumed in central China.

6.2.2.3. Integrated operations of the cascade hydropower complex

As they are hydraulically connected, the joint operations of the two dams require the TGP-GZB to be operated by one entity.

The reservoir starts water storage on September 15, reaching its normal water level by the end of October. The reservoir generally operates according to the operating chart. The GZB primarily plays a reverse regulation role for the TGP. When the TGP undertakes the role of peak and valley regulation for power grids, the GZB helps stabilize the water level in the lower reaches of the river and ensures that the water level downstream is no lower than a certain level to facilitate navigation.

6.2.2.4. Flood control operations

Playing an essential role in the flood control of the middle and lower reaches of the Yangtze River, the TGP is operated by different divisions according to the inflow. There are two types of flood control operation methods, one is normal operating method for ensuring safety of downstream watersheds and the other is for ensuring safety of the hydropower complex.

6.2.2.5. Navigation operations

During the dry season, the TGP operates in a manner that ensures the guaranteed output so that navigation in the river downstream from the GZB can have the required flow level. The current average annual amount of freight is five times that of the highest annual amount before the completion of the TGP.

6.2.2.6. Regulation of sediment

Since the 1990s, the sediment flux to the TGP has steadily declined. Investigations and analyses indicate that the significant reduction of sediment flux to the TGP is primarily attributable to the construction of water conservancy facilities upstream, the execution of water and soil conservation projects, and sand quarrying.

To tackle the potential sedimentation issue of the TGP, the TGP adopts the method of "store the clear and flush the sediment". The amount of water and sediment flowing from the upstream section of the river into the Three Gorges Reservoir is distributed very unevenly over the year. As such, during each year's flood season, the water level of the reservoir is maintained at the flood-control limitation level, thus allowing sandy floodwaters (commonly known as muddy water) to be discharged smoothly downstream. In October after the flood season, the sediment flux to the TGP declines and the reservoir starts storing water to generate power and facilitate navigation.

In order to reduce sediment in the approach channel to the ship locks and to facilitate desilting, the engineering method of "navigation in still water and sand flushing in moving water" has been introduced. With this method, most sediment in the approach channels is washed away, with

the remainder mechanically removed. During the flood season, when the GZB has surplus water, sediment discharging bottom holes and sediment discharging caverns are activated at the right time to discharge sediment.

6.2.2.7. Water replenishment to mitigate droughts

In especially dry years, the TGP may be called upon to replenish water and operate according to the instructions of relative authorities. The reservoir will change the operational focus from ecological protection, navigation facilitation and electricity supply to the power grid to easing the drought, thus significantly easing the shortage of water for human and animal consumption and for farmland irrigation in the middle and lower reaches of the river.

6.2.2.8. Ecological dispatching

The ecological regulation of the TGP refers to the artificial raising of water level in the middle and lower reaches of the Yangtze River by simulating the natural water rise process in an attempt to stimulate the spawning of the "four major Chinese carps" in the middle and lower reaches of the Yangtze River.

6.2.3. Safety and emergency management

6.2.3.1. Dam safety monitoring

Dam safety monitoring includes the monitoring of structure deformation, seepage, stress-strain, hydraulics, strong seismic activities. Structures under monitoring include the right-bank and left-bank dam sections, the guard dam, the underground powerhouse, the ship lock, the reservoir bank near the dam, and the engineering slopes.

6.2.3.2. Cascade complex emergency management

In the company, there is a complete emergency response organization to cover almost any potential crisis. A 24/7 office is responsible for receiving, disseminating and reporting crisis information when it occurs.

6.2.4. Conclusions

The Three Gorges-Gezhouba Cascade Complex is a typical comprehensive case of integrated operation of hydropower stations and reservoirs. It is the biggest hydropower project in the world and brings enormous economic and social benefits mainly including flood control, power generation and navigation.

After years of operation, the CTG has accumulated valuable experience and established a highly efficient operating system to accomplish its multi-target operation and maximize the value of water. The CTG now frequently communicates with other countries' power corporations and it will play a more important role in the development of hydropower around the world.

6.3. ARIÈGE CASCADES IN FRANCE

6.3.1. The context of energy production and water resources in France

Hydropower was the first source of renewable energy in France. Its Installed capacity is 25,400 MW in 2011, ranking second in Europe.

In France, hydropower represents 80% of total renewable energy and about 10% of the total electricity production in 2011. But these figures are rapidly changing with the development of wind and photovoltaic production. In 2011 there were 6,640 MW of installed wind capacities and 2,230 MW of photovoltaic capacities.

Hydropower greatly contributes to reaching the objective fixed by the French government: to produce 23% of electricity from renewable energy by 2020. In 2011, this proportion was 12.8% due to a very dry year (it was of 15.1% in 2010). Hydropower is a high-performing tool for ensuring equilibrium between consumption and production during high levels of electricity demand: hydropower plants can deliver their energy within a few minutes.

Electricité de France (EDF) produces about 8% of its power from hydropower plants. Hydraulics and nuclear enable savings of around 13 million tons in fuel a year in France, strongly reducing CO_2 emissions. There are 435 hydropower plants and 622 dams at EDF, for a total capacity of 25, 400 MW. These plants are located in five geographic regions: the East, the Alps, the Mediterranean, the Pyrenees (Sud Ouest) and central France. The annual production was 34.5 TWh in 2012 for hydropower.

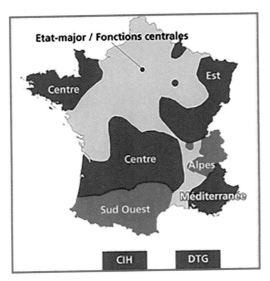

Fig. 6.2
Organisation of EDF hydraulic production

6.3.2. Geography and hydrology of the Ariège Cascade

The Ariège Cascade is located in the Pyrenees region. It takes its source at 2,400m altitude, at Font-Nègre circus, on the border between Andorre and France (Pyrénées-Orientales department). It contributes to the river Garonne on its right bank at Portet-sur-Garonne, situated in the south of Toulouse after a 163 km run from its source.

Ariège has a mean flow rate of 76 m³/s at Pinsaguel for a 4,120 km² basin (situated just above Portet-sur-Garonne), very near from its confluence with Garonne River. The minimum flow rate which measures are taken to sustain the flow of river Garonne is 20m³/s, measured at la Magistère, downstream of Toulouse. This is done principally from the Ariège valley that follows a mountain regime: high level during the snow-melting period from April to June, between 113 m³/s and 156 m³/s, with lower levels outside of this period, and a low level between August and September each year.

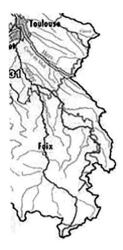

Fig. 6.3
Ariege valley

Hydropower developed very early in the Pyrenees, especially in Ariège because of the steepness of the valley. In 1910, the Orlu Hydropower Power Station was the most powerful hydropower plant in the world, with a total head of 950m.

6.3.3. *Water resource needs*

From its source to its confluence with river Garonne, 17 hydropower plants ranging from 160 MW to a few kW were constructed between 1899 (le Castelet) and 1985 (Ferrières and Laparan), two of which on the Aston tributary, on the left bank, Laparan and Aston, produced about 392 GWh each year – the largest plant production in the Pyrenees region.

Of course this valley is also concerned with many other activities linked to water resources: agricultural needs, drinking water, fishing and fish breeding, recreational activities (canyoning, boating, kayaking), artificial snow, and finally tourism in general.

6.3.4. *Meteorology*

Energy production is weather dependant; therefore since the creation of the EDF in 1946, a specialized unit has been in charge of meteorological predictions (called 'Division Technique Générale' DTG).

These predictions help the producer to manage different risks. About 20 people work at EDF DTG in this field, mainly to maintain the meteorological networks, control the measurements, and apply forecast models, and so on.

EDF has data going back to 1950, including flow rates, height of rain, height of snow, temperature of the air and water, chemical composition of water or fluxes of suspended matter, from about 350 flow rate stations, 400 rainfall stations, 160 snow sticks and 35 NRC (a system that measures the height of snow and the cosmic radiation). All of these data are stored in a system called 'Castor', which can be reached at any time through the EDF network.

6.3.4.1. Short-term forecasts

Teams in charge of predictions may be called upon 24 hours a day, especially for telephone assistance. The aim of this organisation is to forecast extreme events such as floods or very low flow rates.

6.3.4.2. Mean-term forecast

Production forecasts concern about 31 rivers and more than 100 key points; predictions of flow rates, temperature of water and solid transportation are made each day, from date J to J+6.

6.3.4.3. Long-term forecast

Probabilistic predictions of water inflows, from several weeks to several months are performed for forty reservoirs.

6.3.5. Production optimization

6.3.5.1. Context

An electricity market has been required by law since 2005.

The transportation and production of electricity are kept completely separate. RTE (Réseau de Transport de l'Electricité) is in charge of the transportation network (U>42 kV). A contract must be signed between RTE and any producer to access the electricity network. Every producer must ensure equilibrium between its production and the consumption of its clients. EDF ensures its own equilibrium. Each day it must establish a production program for the next day around 4:30 pm, detailed to each half hour. RTE controls the total production program that the net can afford.

6.3.5.2. The reference technical framework (Directive Technique de Référence, DTR)

It is the reference to contact a new plant to the national electricity network. It defines the power that may be fed into or taken from the electricity network and the associated prices, as well as the limit of the network.

6.3.5.3. Service system contract

It is signed between RTE and any party responsible for equilibrium. It concerns adjustments for both frequency and voltage.

6.3.5.4. Optimisation of the production

At EDF, a division is in charge of optimizing the placement of production groups: Division of Optimisation Amont Aval Trading, DOAAT.

It is based on the cost of each production plant. For hydraulic, it is based on the value of the water for each group, which depends on the time and situation in the reservoir; there is a model to determine these water cost values.

6.3.5.5. The production program

Then the production of a group is: Pco+NPr+k (frequency-50Hz), where Pco is the scheduled power production of the group, without secondary frequency regulation and a frequency of the net equal to 50 Hz. Pr is the maximum power absorbed or produced for a step of secondary frequency regulation (N level) and N is the secondary frequency regulation sent by RTE and k the automatic primary frequency regulation.

6.3.5.6. Optimisation of the production program

Based on a statistical approach and meteorological prediction of air temperature, RTE establishes a prediction for consumption.

6.3.5.7. Ariège Production

Large groups are linked to the main electricity network, they are programmed by DOAAT, the programs are sent to one of the three operating centres, for Ariège it is attached to the Lyon centre. The two others are in Kembs in the east of France, the first power plant on the river Rheine, the second near Marseille for plants situated in the south of France, for hydraulic especially the Durance River. Other plants are programmed by specific organisations at regional or local level according to the program established by DOAAT.

6.3.6. Safety management

These situations are treated case by case, but in general, there are structures dedicated to:

- First, a 24/7 organisation to deal with all sorts of issues at the level of several hydropower plants (Group of hydraulic plants: GEH Groupe d'Exploitation Hydraulique)

- A crisis cell may be constituted from the management of the GEH; it may be activated in any crisis situation. Well-trained teams will manage these situations.

- As described in chapter 3, DTG and its meteorological prevision cell, give indications of the risk of flooding each day and produce more information when there are floods.

- Risk analysis is done along every hydraulic site, upstream and downstream of the plant.

As for seismic events, an outside structure of the EDF sends messages to the command (PHV: Poste Hydraulique de Vallée).

6.3.7. Conclusions

Ariège is a good example of a water resource that has to be shared between several economic actors, especially agriculture and recreational activities.

Historically, hydropower developed in Ariège at the beginning of the 20th century. Nowadays satisfaction of economic actors, preservation of the environment and production of renewable, and carbon free energy have to be managed. The law on aquatic fields gives the framework and organisation to discuss these issues. It is complex but also necessary for a lasting solution to the treatment of our water resources.

Hydropower production is susceptible particularly to meteorological hazards. Specific organisations have to manage any crisis arising from meteorological hazards or from any other source of hazard such as seismic events.

6.4. THE DEZ RIVER SYSTEM IN IRAN

6.4.1. Introduction

The major hydropower plants in Iran, generating firm energy, belong to the Great Karun River Basin, which is located in the south west of Iran with an area of 67,257 square kilometres. The mountainous part of the basin is about 45,994 square kilometres (68 percent) and the rest of 32 percent (21,263 square kilometres) is foothill and flat plains. The Great Karun River Basin is formed from two main sub basins: the Karun, and the Dez River basins, with a total of 12 hydropower dams (5 are under operation, 4 are under construction, and 3 are under study). The Dez sub basin has been selected as a case study for a pilot project for integrated water resources management with hydropower generation as its primary objective.

The Dez River, itself, consists of two sub basins, and when they are confluence, they form the Dez River and shortly enter the Dez dam's reservoir. At present, the power plant at the Dez dam includes 8 units with a total capacity of 520 MW. It is intended to rehabilitate the existing unit from 2013 and prospected by end of year 2016 to increase the capacity of each unit to 90 MW and totally 720 MW. The second unit of the Dez power plant is expected to be ready for operation by 2017, with a capacity of 4*180 MW (with existing rehabilitated units, it will be 1440 MW in total).

The water is regulated in the Dez reservoir for energy generation, water supply for downstream water demands, and flood control. The Dez River finally joins the Karun River at the Band-e-Ghir junction to form the Greater Karun River.

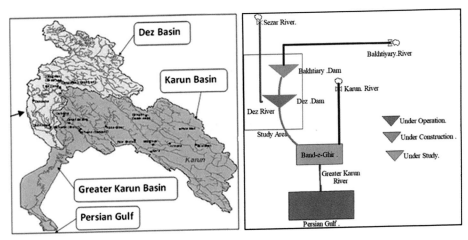

Fig. 6.4
Dez River System

6.4.2. Multipurpose operation

6.4.2.1. Water demand

(1) Environmental water demand

The environmental demands on the Greater Karun River are for water quality control and prevention of Persian Gulf saltwater intrusion. The environmental water supply for the Dez River is calculated to be about 930 million cubic meters (30 cubic meters per second), which accounts for 90 percent of the whole environmental water supply.

(2) Potable, Industrial and Irrigation water demand

Consumptive water demands for the study area include potable, industrial and irrigation water demands downstream of the Dez dam. The consumptive demands upstream and downstream of the Dez Dam taken in this study are as follow:

- The monthly demand downstream

The monthly demand downstream of Dez Dam can be separated in two major parts: demand from the Dez and the demand from diversion weir.

- Future water resources projects

Based on the official statement, the future water resources development in the Sezar and Bakhtaran sub basin is defined, which are mainly used for irrigation.

- Inter basin water transfer

The Inter basin water transfer is proposed to transfer water from upstream of Bakhtiary sub-basin to the Qom-rud basin.

6.4.2.2. Sediment Dispatching

The total annual suspended sediment inflow to the Dez reservoir is estimated to be 14.24 million tonnes with a bed load of 3.56 million tons per year. Most of the sediment enters the Dam during the wet season in April, May, and June. Due to the problem of sediment deposition in the Dez reservoir, part of the sediment from reservoir was attempted to be discharged by flushing.

6.4.2.3. Flood control

One of the purposes of the Dez dam reservoir is flood control. The normal water level during non-flood season (middle of April till the end of October) is 352 masl, while during flood season (November to middle of April) the water level drops to 337 masl. The flood control volume is designed to control 100 years of flooding.

6.4.2.4. Reservoir operating rules to supply minimum energy during drought period

The main objective of the Dez dam Reservoir is to generate firm energy during peak power demand. Therefore, energy supply with acceptable reliability is crucial. In order to fulfil this objective, a reservoir rule curve is designed for energy generation during drought period to improve energy generation in drought conditions. Also, a hedging rule has been designed to reduce the energy generation with different coefficients proportional to the reservoir volume.

6.4.3. Conclusions

Based on the objectives of the Dez Dam, the projects which are going to be implemented in this basin can be categorized as follows: small projects under construction, inter-basin water transfer, large projects, and future projects (Dez Dam heightening, upgrading the existing power plant, etc.). These projects may have an effect on water resource potential and different scenarios are carried out to analyse the effects on water resources. The fact that can be drawn from the conclusions is that following projects will reduce water regulation of the Dez Dam for energy generation and water supply downstream of Dez Dam and have negative effects on downstream water consumption.

6.5. KISO RIVER STUDY IN JAPAN

6.5.1. Introduction

As one of the largest rivers in Japan, The Kiso River is 229 km long originating from Mt. Hachimori in Nagano Prefecture and covering an area of 5,275 km² with abundant water resources which is valuable to agriculture and various industries (Figure 6.5). It is managed by the Kansai Electric Power Corporation (KEPCO) which operates 151 hydropower plants with a total output of 8 GW, of which more than half are pumped-storage power plants.

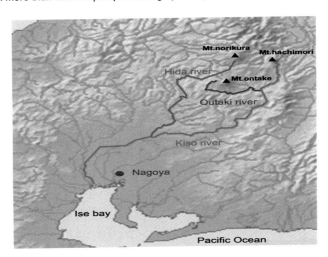

Fig. 6.5
The Kiso River basin

In the 1910s, hydropower development of the Kiso River system started to meet growing demands for power, and nowadays almost all the head and flow of the river is utilized to meet the demand for power. With rapid urbanization driven by high economic growth, demands for waterworks, industrial water and flood control increased on the downstream plain areas where populations and industry are concentrated. As a result, construction of multipurpose dams (Figure 6.6) was promoted to unify flood control and water usage.

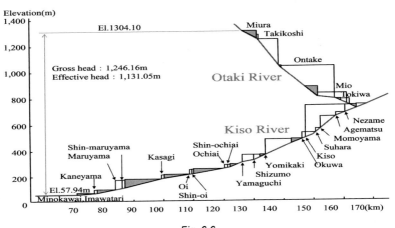

Fig. 6.6
Longitudinal Profile of the Kiso River System and Hydropower Facilities

There are 13 dams and 33 power plants on The Kiso River and its tributaries, with total output accounting for more than 1,000MW. All the power generated is transmitted to the Kansai region or large power usage areas. Meanwhile, dams are broadly classified into water use dams and flood control dams according to their purpose, and water use dams are further categorized into irrigation dams, waterworks dams, industrial dams and hydropower dams. And multipurpose dams are also built on the same river system, which are used for two or more of the above purposes.

6.5.2. Comprehensive operation and dispatching

6.5.2.1. Framework for generating operations

Figure 3.3 shows the chain command for related units of the generating operations of the hydropower facilities on the Kiso River, as well as their major duties.

(1) Central Load Dispatching Center (CLDC)

The objectives of the Central Load Dispatching Center (CLDC) are to supervise and control a balance between power supply and demand in the entire service area of KEPCO, and to make a request for generating operation to each Local Load Dispatching and Control Center (LDCC).

(2) Tokai Load Dispatching and Control Center (Tokai LDCC)

Tokai LDCC, one of KEPCO's regional load dispatching offices, can remotely start and stop all the generating units in response to dispatching orders from the CLDC. Depending on the supply and demand, weekly and daily generation schedules for each power plant are drawn up by Tokai LDCC in cooperation with the CLDC.

(3) Dam Control Office

Dam operators are on duty of twenty-four-hour shifts at each dam with spillway gates. Beside flood control, their major role in generating operations is to handle power intake gates at open-channel types of power plant in response to dispatching requests from the Tokai LDCC.

(4) Power System Division

The annual power demand forecast and supply plan for KEPCO's service area are studied by the power system divisions in the headquarters. They prepare the annual reservoir operation plan for the Miura Dam, which is located upstream of the river, and then submit it to CLDC and LDCC for their generation schedule.

6.5.2.2. Major considerations for the generation schedule

The major function of the cascade hydropower plants on the Kiso River is to provide power to meet changing demands of the load during peak and middle peak periods and try to maximize the total power generation of the entire river system at the same time.

Tokai LDCC and the CLDC cooperate to draw up annual, monthly and daily generation schedules by considering the state of natural river flow, out-of-service periods due to maintenance work, volume of reserve capacity, types of hydropower generation (reservoir type and regulating pond type) and hydropower plant (pressured tunnel type and open channel type), etc.

6.5.2.3. Generation schedule

(1) Annual supply plan

The Power System Division prepares the annual power demand and supply plan to secure dependable capacity corresponding to the forecasted demand at the beginning of the fiscal year. The annual supply plan is drawn up in order to estimate a dependable amount of power output for this forecasted demand.

(2) Monthly generation schedule

The CLDC draws up the monthly generation schedule for three months, in which they try to select more economical operations to compare the difference in operational costs, and they also incorporate the current status of planned outages to secure dependable capacity and reserve capacity in case of an unexpected accident.

(3) Weekly generation schedule

CLDC studies the weekly supply plan based on weekly demand forecasts in order to provide a stable and economical supply of electricity, and then make a request to Tokai LDCC for generating operations during peak demand time running the specific hydropower plants equipped with AFC and ELD devices. Tokai LDCC studies the weekly generation schedule on an hourly basis for each plant to take into account fluctuations of natural river flow while referring to weekly weather forecasts, as well as the reservoir operation plans of the Miura Dam.

(4) Daily generation schedule

The CLDC forecasts the power demand for the next day referring to the latest weather forecast and reviews the details of the daily supply plan by checking the changes in supply power, reserve capacity, reliability on the grid system etc. Tokai CLDC reviews the weekly generation schedule and then determines the final hourly generation schedule for each plant. Tokai LDCC then operates their power plant as scheduled the next day and makes requests for power intake operations to dam control offices.

6.5.2.4. River flow forecasting

(1) River flow forecasting for the supply plan and generation schedule

River flow or weather forecasting are not applied in the drawing up of the annual supply plan and monthly generation schedule, both of which are based on actual inflow data over 30 years. Tokai LDCC studies and reviews the weekly generation schedule as well as the reservoir operation plan by using weekly weather forecasts distributed by the Japan Meteorological Agency, considering such conditions as atmospheric pressure distribution, location of a cold front, rain area, etc.

Weather forecasts sometimes prove to be wrong and thus it is not predicted to rain on the next day, the river flow for the next day is deducted by a certain amount, which is empirically estimated or based on accumulated data, from the river flow used in the present day's generation schedule.

(2) River Flow Forecasting at the Dam Control Offices

There are two types of river flow forecasting system that are currently in use: one is that runoff extending for 10-minute, 30-minute, one-hour, two-hour and 3-hour periods are predicted with the storage function method as a run-off model. The other is a hybrid rainfall prediction model that is composed of extrapolation model using radar data and meso-scale atmospheric model, which can predict six hours ahead at an interval of 10 minutes. This model is combined with a distributed runoff prediction model to constitute the real-time dam inflow prediction system.

6.5.2.5. Frequency regulation

As generating units of hydropower plants have a higher ability to response to sudden demand changes than other types of unit, several hydropower plants on the Kiso River are equipped with AFC, ELD, and Governor-free devices for ancillary services in the service areas.

6.5.3. Other considerations

6.5.3.1. Restrictions on the storage of water

There is a re-regulation dam, the Imawatari Dam is located at the point most downstream in order to temporally store the fluctuating river flow due to power generation during peak period and then evenly release the water for the downstream. For this purpose, the storage of water at each dam upstream from the Imawatari Dam is restricted when the discharge from it falls below 100 m³/s on a daily average basis.

6.5.3.2. River basic flow

In order to properly control river comprehensively in the low-water period, the discharge necessary shall be specified for maintaining regular functions of river flow at a major point on a river.

6.5.3.3. Renewal of water rights

It is indispensable for the power utilities to get permission for the operation of hydropower plants. The renewal of water rights is also needed at a certain interval.

6.5.3.4. Risks in transmitting electricity to the Kansai region

All the generated power is transmitted through 275km transmission lines to the Kansai region more than 200 km away. In order to reduce such risks that would lead to the decrease in expected supply power, there are three transmission lines.

6.5.4. Emergency management

6.5.4.1. Power interchange

There are two types of frequency in Japan: all electricity in west Japan is 60Hz, while that in the east is 50 Hz. Japan has three frequency converter stations, which can convert 50 Hz into 60 Hz and vice versa, but their capacity is quite limited.

6.5.4.2. Drought

Drought conciliation is one kind of management of water use in an emergency. In accordance with river laws, there is an agreement among the parties concerned to establish a special purpose committee for coordination when industrial water, irrigation, waterworks, etc. are impacted by drought on the Kiso River.

6.6. COMPREHENSIVE COUNTERMEASURES AGAINST THE RECORD-BREAKING DROUGHT IN KOREA

6.6.1. Background of drought in 2015

In 2015, rainfalls and water levels of major dams in Korea marked the lowest record in history, which was described as the record breaking drought ever since the beginning of official hydrological observation. 20-year frequency drought inflow has been normally used for designing the dam storage volume for water supply. Most of inflows into dams were smaller than 20-year frequency drought inflow. That means 2015 case can be classified into natural disaster. Continuous drought condition since 2014 threatened normal dam operation.

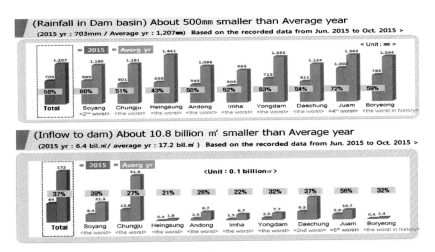

Fig. 6.7
Rainfall and inflow in Major dams in Korea in 2015

6.6.2. Anticipatory actions to cope with worsening drought

In order to efficiently cope with the drought, Enterprise Drought Task Force Team in K-water had been composed and implemented since July of 2014. Furthermore, a water supply adjustment criterion in time of drought was established by K-water. It has been applied to dam operation after the government approval. The relevant stakeholder including dam operators easily recognizes the current drought stage by this criterion because this provides the particular standard water levels around year and also step by step water supply adjustment plan according to drought stage. As a result of this criterion, fast decision making such as reduce of water supply has become possible.

There are four sorts of water supply from a multi-purposed dam, it would be classified into domestic, industrial, irrigational and river maintenance water supply according the priority as presented in Table 6.1. Drought response stage can be classified into 4 stages in accordance with water demand, namely notice, caution, alert and serious.

Table 6.1
Explanation of standard storage volume in stages

Stage	Estimation Criteria in Stages
Notice	Standard storage volume for Caution stage+ Possible to supply water demand 10 days without inflow
Caution	Possible to supply actual water demand for 1 year (Demand: river maintenance, Irrigation, Domestic, Industrial)
Alert	Possible to supply actual water demand for 1 year (Demand: Irrigation, Domestic, Industrial)
Serious	Possible to supply actual water demand for 1 year (Demand: Domestic, Industrial)

K-water made an effort to preferentially secure the domestic, industrial and irrigational water by reducing river maintenance water because three kinds of water supply directly relate to state of the economy and the public daily life.

6.6.3. Additional countermeasure in time of drought worsening season

Despite of anticipatory actions, various counter measure were carried out as the drought condition spread out over the whole country and the situation getting worse. In Han river basin which is the main water source of Seoul metropolitan, hydropower dam substituted domestic water supply instead of multipurpose dam due to gradual drop of multipurpose dam water level. Furthermore, absolutely required water was only provided by joint investigation with government ministry and relevant local government for the actual downstream extract.

In Nakdong and Geum river basin, irrigation water from dams had been supplied actual demand since September of 2015. And dam water release had minimized while maintaining the water level of downstream without interruption of extracting water by conjunctive operation among dams, weirs and barrage. In Sumjin river basin, the hydropower dam which used to produce power generation by diverting water to other area with high head had changed its release direction to mainstream in order to fill the downstream dam storage volume.

In Boryeong dam, the main water source of 8 southwest local governments, reservoir water level gradually reached to the low water level in spite of anticipatory actions. In order to prevent stopping of water supply from dam, various and urgent projects had been implemented as followings.

Supply Management (Structural)	• **Alternative supply connecting with different multi-regional water** supply system • **Urgent leakage reduction:** water flow rate 10% ↑ within 6 months • Construction of Water Supply diversion canal **(Geum river→Boryeong dam)** : 115×10³m³/day

Demand Management (Non- structural)	• **Autonomous adjustment for water supply:** water usage 20% ↓ without compulsory cut off • **Support fund for water saving:** nation's first, inducement of water saving by incentive pay • **Water Saving Campaign:** mass media(including press tour), road campaign, etc. • **24hr call center for civil complaint, Support bottled water and water wagon**

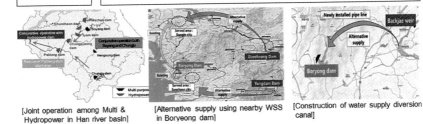

[Joint operation among Multi & Hydropower in Han river basin]   [Alternative supply using nearby WSS in Boryeong dam]   [Construction of water supply diversion canal]

Fig. 6.8
Additional countermeasure in time of drought worsening season

6.6.4. Successful outcomes by various actions in spite of historical drought

Although unprecedented extreme drought spread out over the whole country, K-water has been stably provided required water to the public by means of comprehensive countermeasure. With anticipatory and active actions against drought, additional storage volume(2.4×109 m³) among 9 dams was secured and drought stage would be mitigated.

Table 6.2
Additional reserved volume and drought stage

Description	SY	CJ	HS	AD	IH	YD	DC	JA	BR	Total
Practical action stage (with water saving)	Caution	Caution	Caution	Caution	Caution	Caution	Alert	Caution	Serious	-
Without action coping with drought (without water saving)	Under Low Water Level					Caution	Alert	Caution	Serious	-
	Serious	Serious	Serious	Serious	Serious					
Additionally reserved Storage Volume [10⁶m³]	1,890		23.2	345		13	90	35	10.5	2,405

* SY : Soyanggang dam, CJ : Chungju dam, HS : HoengSung dam, AD : Andong dam, IH : Imha dam, YD : Yongdam dam, DC : Daecheong dam, JA : Juam dam, BR : Boryeong dam

6.7. CASCADE HYDROPOWER RESERVOIRS IN NIGERIA

6.7.1. Introduction

Nigeria has huge power supply deficit. Whereas the Power demand (Power Sector Reform Roadmap, 2010) is about 40,000MW, the total installed capacity is about 4,000MW. The Hydropower plants contribute about 1,900MW.

The Nigerian Cascade Hydropower case study presented here involves three main Cascade Reservoirs within the Hydropower generation complex. These are Kainji-Jebba cascade with a capacity of 1,400MW in River Niger, Shiroro-Zungeru cascades with a capacity of 1,300MW on River Kaduna and Gurara I – Gurara II cascade with a capacity of 390MW on River Gurara. Only one of these three (Kainji-Jebba) is fully in production. The Shiroro-Zungeru and Gurara I & II cascades will be in production by year 2018 based on current developments and program.

Fig. 6.9
Three HPP Cascadesin Nigeria

6.7.2. Meteorological and runoff Characteristics

The Climate is characterized by marked dry and wet seasons and annual rainfall varies from about 2,500mm in the extreme South to less than 500mm in the extreme North. The Kainji-Jebba Hydropower plants operating on River Niger enjoys better hydrological performance in any given year

than the Shiroro-Zungeru cascade on River Kaduna because the former exhibits two flood peaks in September and February (See Fig6–10a and Fig6–10b). The River Kaduna cascade enjoys only three months of flood flow between July and October annually while River Niger at Kainji experiences two annual floods. The Gurara, though a smaller tributary of River Niger shows similar flood regime to River Kaduna. In general, the three cascades show marked seasonal variation.

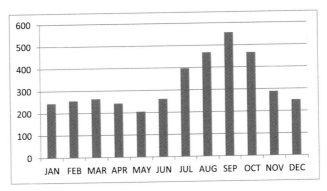

Fig. 6.10a
Average Monthly Runoff of River Kaduna Mean Discharge (m³/s)

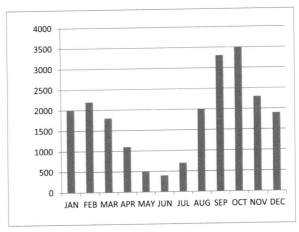

Fig. 6.10b
Typical Hydrograph of River Niger Showing Average Monthly Runoff
Discharge (m³/s) with two Flood Peaks

6.7.3. *Comprehensive operation and dispatching*

The operators employ water level records without forecasting in three cascades. Hydropower plants in the three river cascades are used mainly for hydropower generation, flood control, sediment dispatching and are available for navigation. The effective management of the reservoirs in response to inflow flood allows high energy generation for certain months.

6.7.4. Conclusions

The scope of the existing practice in Nigeria is very limited and rudimentary. Power generation companies do not have facilities for real time forecasting of relevant hydrological data. The new Investors of the River Niger and River Kaduna Cascades are investing significantly in the rehabilitation of the hydropower plants including technology for efficient operation and dispatching of the hydropower reservoirs.

6.8. CASCADE HYDROPOWER STATIONS ON THE VOLGA RIVER AND KAMA RIVER IN RUSSIA

6.8.1. Introduction

The Volga River is the largest river in Europe located in the European part of Russia and. It is 3530km long originating from a spring in the Valday Hills in Tver Region and covers a catchment area of 1.36 million km^2 with about 200 tributaries of which the biggest are the Kama-river and Oka-river. Cyclones from the Mediterranean Sea bring heavy rains in summer and thaw it in winter, so the Volga has an Eastern European type of water regime with spring floods (April-June), low summer and winter mean inflows and autumn rain floods (October).

There are 9 large reservoirs on the Volga River and 3 large reservoirs on the Kama River. The longitudinal profile of the Volga and Kama river system and hydropower facilities is given in Figure 6.11.

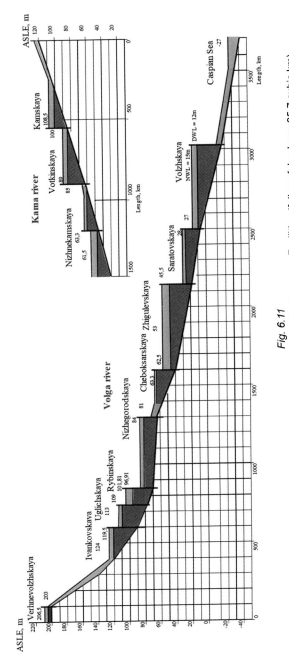

Fig. 6.11
Longitudinal profile of the Volga and Kama river systems and hydropower Facilities (full useful volume 85.7 cubic km)

6.8.2. Comprehensive operation

All reservoirs of the cascade are used complexly and creation of Volga-Kama cascade of reservoirs allowed to solve a number of issues for energy, water-transport and irrigation development, as well as to provide industry and utilities with water supply.

The Upper Volga and Kama are energy and water transport sectors, the main function of Ivankovskoye Reservoir is water supply to Moscow (it provides up to 70% of water consumed in the city). On the Middle Volga, irrigation is added to the main demands of the water system, and on Kama River there is timber rafting. On the Lower Volga, in addition to energy and water transport, leading sectors are fisheries (fishery releases) and agriculture (agricultural releases and irrigation).

The objectives of Volga operations are identified as follows: drought management for the salvation of eastern Volga areas from periodic crop failures; ensuring guaranteed grain production through irrigation development; development of electric power through construction of powerful HPP cascades on the rivers Volga and Kama; development of a unified water-transport system that would allow large vessels to transport goods and passengers.

The Volga together with its tributaries are also the source of potable and industrial water for the population and economy of the Volga basin. Potable water extraction from the Volga is about 26 km³ per year. Irrevocable water consumption is close to 10 km³ per year. The use for household need is 29.2% of total extraction, 51.4% for industry, 9.1% for irrigation and the rest 10.3% for other purposes.

6.8.3. Dispatching schedules

6.8.3.1. Characteristic lines and zones

In accordance with accepted water practices, the volume of the reservoir is divided into the following specific areas of dispatching schedule:

- Zone of unused volume of the reservoir, which is located below the minimum acceptable in normal use level;

- Interruption zone (reduced return), in which the return of the reservoir is assigned below the guaranteed level;

- Zone of guaranteed return, which is the main work area, where guaranteed return is appointed;

- Zone of high (excess) returns over the guaranteed level. This increase in returns provides the additional effect mainly to hydropower;

- Flood zone prism occurs in reservoirs used for the protection of the downstream hydroelectric facility against flood;

- Zone of maximum idle discharges. In order to ensure the safety of hydraulic structures all spillways are open in this area.

It should be emphasized that the above specific zones change their position in the dispatch schedule, depending on the characteristics of the water phase in any year. In addition, the boundaries of these zones at certain moments and time periods may vary depending on the preceding or predicted values of inflow. The main importance to the dispatch schedule is characteristic lines separating zones and subzones. On these lines, there is a change in management strategy of reservoirs.

6.8.3.2. Algorithm for dispatching schedule

The optimisation algorithm of water energy regimes for reservoirs cascade is presented in the following table.

Table 6.3
Optimizing the water regimes

Title	Long-term planning	Medium-term planning	Short-term planning	Executive planning
Key procedure	Forecasting balance sheet making	Making proposals to inter-authority working group and water discharge calculation after receiving the directives	Making a schedule for the following day and week	Choosing a set of units to perform power schedule
Effect	none	Amplifying the output by discharge before flood period, dispatching of discharge between HPPs and periods using price criteria	Choosing the optimal schedule for avoiding idle discharge, for price criteria and minimizing limitations	Minimizing water discharge
Actor	Federal tariff service, Power forecasting agency	Federal water resources agency	Transmission system operator	Transmission system operator, RusHydro

6.8.3.3. Dispatch operating process

Based on characteristic lines and zones of dispatch schedules, in general, dispatch operating process includes:

- Products: Energy (power supply schedule compliance) and power (declared available power providing);

- System Services: Reservoir level and discharge, voltage level and frequency level in the grid(power supply reserve, and command execution to regulate voltage and regulate supplementary frequency and power flow);

- Transmission lines operator – sets up power supply schedule in terms of delivery of the available power, give directives to regulate voltage, frequency and power flow. Federal water resources agency – sets up water discharge schedule, send directives on water level operating range and discharge;

- Consumers: Wholesale market – consumes energy and power.

Grid operator and Federal Transmission Lines Company consume voltage, frequency and power exchange.

Water consumers – consume volume of water usage.

6.8.4. Limitations and goals of dispatching schedules

When planning the water and power regimes of the Volga River and Kama River cascades, it is necessary to take care of specific features and special aspects of limitations for their operating regimes:

- Instructions made by authorities;
- Climate in the region of watershed basin and in tail-water;
- Geology and morphometry of reservoir area, power site and tail-water;
- Dam safety;
- Power mix in the energy sector and patterns of energy consumption;
- Type of flow regulating at the reservoir (daily, weekly, monthly, seasonal, long-term);
- Requirements of water users and customers for the water regime.

Based on these limitations and relevant parameters, the goals when planning water and power regimes are:

- Safety and reliability of HPP's operating equipment;
- Dam safety;
- Basing the position of company when drawdown and impoundment of reservoir regime is set by the basin authority;
- Maximizing power output or maximizing sales proceeds through optimal water and power regime, taking into account the limitations set by authorities.

6.8.5. Conclusions

The operational management of reservoirs was developed on the basis of an accumulation of experience in the operation of reservoirs, improvements in scientific approaches to the regulation of reservoir flows, and the emergence of new challenges and requirements for reservoir regimes. The rapid development of computer technology and mathematical modelling methods has also exerted a great influence.

6.9. THE UPPER RHONE RIVER VALLEY IN SWITZERLAND

6.9.1. Introduction

The Rhone River, originating in the southern Alps, is 812 kilometres long and covers a catchment area of 97,000 km². The development of hydropower in the Alps, and especially in the upper Rhone river valley, started towards the end of the 19th century and the beginning of the 20th century. Investors took advantage of the proximity of electrical power generation of to developing industries such as chemical, aluminium and manufacturing. The first hydropower schemes mostly consisted of a water intake located in a lateral valley, possibly combined with a small reservoir, and high chute penstock ending in a powerhouse located in the main valley. With the increasing demand for hydropower, high reservoirs located in more remote upper lateral valleys were constructed, mostly

between 1945 and 1980. In order to artificially increase the catchment areas of the major schemes, an important network of water intakes and collecting water galleries was constructed. The total catchment area controlled by high reservoirs represented 27% of the global watershed (5,500 km²).

Nowadays, the total retention volume of the reservoirs is about 1,200 million m³ representing 21% of the annual discharge flowing into the Lake of Geneva. Globally, 85% of water is used, at least once, for power production.

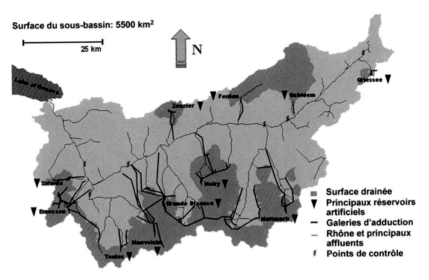

Surface du sous-bassin: 5500 km²

Surface drainée
▼ Principaux réservoirs artificiels
— Galeries d'adduction
— Rhône et principaux affluents
ƒ Points de contrôle

Fig. 6.12
Map of the Upper-Rhone watershed (5500 km²)

Although the most important floods appear in September and October, when the hydropower reservoirs are almost full, these reservoirs offer the potential for reducing the impact of flooding events in the main Rhone river valley.

6.9.2. Improving the effect of flood routing

A first project aiming at improving the flood routing effect of an existing reservoir was developed in 2001 for the Mattmark dam located close to Zermatt and the Matterhorn. An additional retention volume was created by heightening the existing spillway by 2 metres, the normal operating level remaining the same as before. The original freeboard of 7 metres has been reduced to 5 metres without endangering the dam itself. The additional volume in the reservoir contributes therefore to the flood routing.

Nevertheless, it has also been demonstrated that the impact on existing reservoirs and possible improvements of their flood routing effect are beneficial for flood protection only for flooding events of medium importance, remaining of very limited use in the event of more extreme flooding. Further developments aimed at integrating flood and hydropower management require flood forecasting and management decisions in real time.

6.9.3. Managing floods taking advantage of the hydropower reservoirs

6.9.3.1. Flood forecasting

The influence of flood routing from hydropower reservoirs depends highly on reservoir management prior during the flooding event. Optimization of the retention effect during floods requires a flood forecasting system and routing simulations in order to provide real time decision support. The use of a discharge forecast system gives the required information about the hydrological situation in the catchment area and provides the decision-maker with a general view and a predicted evolution of the discharges in the river network. When coupled with an optimization tool, it is possible to directly highlight the key variables of the system and to decide which release or storage operations at the reservoirs have to be performed.

Flood forecasts are based on meteorological data including precipitation, evapo-transpiration and temperature, which are provided 72 hours in advance. A semi-distributed hydrological model integrates this data and simulates the snow and ice melting, infiltration and run-off processes. The river network is also modelled as well as the hydraulic schemes and equipment such as reservoirs, water intakes, spillways, power plants, pumps.

The results of the simulation of discharge at a certain number of control points are then compared to those measured in real time. the model is then updated and the new meteorological data introduced for further forecasting over the next 72 hours.

6.9.3.2. Preventive operation

Preventive operations refer to an anticipated reduction in the volume of water in certain reservoirs. The release of water can be performed by power production or, if necessary, by operating gates. Since the decision is based on hydrological forecasting, the risk of loss of power production, or the risk of power generation during a low economical timeframe exists. It is therefore necessary to validate each decision when comparing forecasting and real time measurements.

6.9.3.3. Performance

A performance analysis of the computed flood management strategies showed the significant positive influence of such preventive operations. Two major historical flood events with about 100-year occurrence periods were simulated. The reduction of the observed peak discharge due to the accumulation reservoirs lay between 6% (1993 flood) and 10% (2000 flood) at the outlet of the Rhone River catchment area. With an appropriate flood management strategy 30 hours in advance, the reduction of peak discharge could have been 15% without releasing water by gate operation and 28% with gate operation during the flood in September 1993. An additional reduction of 7% of the peak discharge would have been obtained during the flood in October 2000 with preventive operations 34 hours in advance.

6.10. THE TENNESSEE VALLEY RIVER SYSTEM IN THE USA

6.10.1. Introduction

The Tennessee Valley River is 1,600 km-long, with an area of more than 100,000 km^2. It has abundant precipitation of up to 1,320mm per year. For the use, conservation, and development of water resources related to the Tennessee River, the TVA operates a system of dams and reservoirs with associated facilities – its water control system. TVA uses this system to manage the water

resources for the purposes of navigation, flood control, power production and, consistent with those purposes, for a wide range of other public benefits.

TVA maintains 29 conventional hydroelectric dams throughout the Tennessee River system and one pumped-storage facility for the production of electricity. The total number of regulated dams maintained by TVA River Operations has risen to 87, with the addition of saddle dams, dikes and non-power dams. In addition, four Alcoa dams on the Little Tennessee River and eight U.S. Army Corps of Engineers (USACE) dams on the Cumberland River contribute to the TVA power system.

TVA began automating its hydroelectric system in 1998. The program was completed in 2005, allowing all TVA conventional hydroelectric plants to be controlled and monitored from a command centre in Chattanooga. An ongoing hydro modernization program, scheduled for completion around 2015, will result in an additional 360 megawatts available from existing hydro units.

6.10.2. Background and water control system overview

This chapter describes the seasonal patterns of rainfall and runoff in the Tennessee Valley watershed and the specific components of the TVA water control system.

6.10.2.1. Rainfall and runoff

Rainfall, runoff, and topography in the Tennessee Valley watershed strongly influence the original location, design, and operating characteristics of TVA reservoirs and the water control system. The locations and storage volumes of reservoirs reflect the variation in rainfall and runoff in the region. Rainfall and runoff continue to control when and where water flows into the reservoirs; and runoff exerts a strong influence on the annual, seasonal, and weekly patterns of reservoir operations.

Mean total annual rainfall is 52 inches per year throughout the TVA system, but rainfall varies considerably from year to year and at different locations in the system. About 40 percent of rainfall in the drainage area of the Tennessee River system becomes runoff. Substantial variation in the annual amount of rainfall affects the degree to which objectives of the water control system can be achieved.

6.10.2.2. Structure of the Water Control System

The water control system is composed of dams and reservoirs, tail waters, navigation locks, and hydropower generation facilities, as described in the following sections.

(1) Dams and Reservoirs

The 35 projects that comprise the water control system include nine main stem reservoirs and 26 tributary reservoirs. Each TVA project typically falls into one of four general categories that are closely related to its characteristics (e.g., location and size), primary function (e.g., navigation, storage for flood control, or power generation), and operation. These categories include main stem storage projects, main stem run-of-river projects, tributary storage projects, and tributary run-of-river projects.

(2) Tailwater

Tailwater is a widely used term that generally refers to the portion of a river below a dam that extends downstream to the upper portion of the next reservoir pool in the system.

(3) Navigation Locks

The TVA reservoir system also includes 15 navigation locks located at 10 dams. Operated by the USACE, the locks provide an 800-mile commercial navigation channel. TVA operates the reservoir system to maintain a minimum 11-foot depth in the navigation channel along this navigable waterway.

(4) Hydropower Generation Facilities

Hydropower generation facilities are incorporated into 29 of the project dams. Although these facilities initially provided base load power, they now generate electricity primarily during periods of peak power demand. Depending on annual runoff, the hydropower facilities provide from 10 to 15 percent of TVA's average power requirements.

6.10.3. Water control system

This section describes how the water control system is operated to optimize public benefits while observing physical, operational, and other constraints.

6.10.3.1. Flows through the Water Control System

Water stored in the tributary reservoirs is released downstream to the larger Tennessee River main stem projects and eventually flows into the Ohio River, and finally the Mississippi. Water is released from the projects to provide flows to maintain minimum navigational depth, re-establish flood storage volume in the reservoirs, generate power as it passes through the system, supply cooling water to the coal and nuclear power plants, and maintains water quality and aquatic habitat.

6.10.3.2. Balancing operating objectives

The TVA reservoir system is not operated to maximize a single benefit to the exclusion of others. The system is operated to achieve a number of objectives and to provide multiple public benefits. Some operating objectives are complementary; others require trade-offs, especially in periods of limited water.

6.10.3.3. Reservoir operations policy

TVA's reservoir operations policy establishes a balance of operating objectives. It guides system-wide decisions about how much water is stored in specific reservoirs, how the water is released, and the timing of those releases. The policy helps TVA in managing its reservoir system to fulfil its statutorily prescribed operating objectives and to provide other benefits.

The reservoir operations policy is composed of guidelines that describe how the reservoirs should be operated given the rainfall and runoff and the operating objectives. These guidelines include:

(1) Reservoir operating guidelines

Control the amount of water in each reservoir, the reservoir pool elevations, and the flow of water from one reservoir to another; these guidelines are implemented through guide curves for each reservoir.

(2) Water release guideline

Control the release of water needed for reservoir system and project minimum flows, including flows for special operations.

(3) Other guidelines and operational constraints

Include procedures and limitations set for hydropower generation, response to drought conditions, scheduled maintenance for power generation facilities, power system alerts, dam safety, security threats, and environmental emergencies.

(4) Reservoir guide curves

Guide curves are line graphs showing the planned reservoir levels throughout the year. They also depict the storage allocated for flood control, operating zones and, in some reservoirs, the volume of water available for discretionary uses.

6.10.4. System monitoring and decision support

To ensure the efficient operation of its complex reservoir system, TVA uses a variety of data collection, computerized reporting, and decision support systems.

6.10.4.1. River Forecast Center (RFC) mission

The TVA's RFC is staffed around the clock, 365 days a year. River schedulers continually monitor weather conditions and water quality data, as well as water availability and demand, with the goal of routing water through the river system to provide the most public value given changing weather conditions and water needs. The RFC responsibilities include:

- Issuing forecasts of reservoir levels;

- Scheduling water releases at TVA dams;

- Providing hourly generation schedules for TVA hydroelectric projects, eight projects operated by the U.S. Army Corps of Engineers on the Cumberland River system, and four reservoirs that make up the Brookfield Renewable Energy Smoky Mountain Hydro project;

- Providing special notifications to the public during flood events;

- Evaluating cooling water needs for TVA coal-fired and nuclear plants;

- Monitoring water quality conditions below TVA dams so that aeration equipment can be turned on when needed to maintain adequate dissolved-oxygen concentrations;

- Serving as the main point of contact in the event of a river system emergency.

6.10.4.2. RFC challenges

While the RFC System continues to operate as a functional river forecasting system capable of developing a daily water schedule, it is an aging system that is in need of modernization. The RFC System is poorly suited to meet the intense demands currently placed on the Tennessee River network. Some of the consequences of this include:

- System failure risk due to thinness of support knowledge and lack of documentation.

- Inflexibility due to custom connections between components.

- Outmoded tools due to cost of keeping current with custom software.

- Waste of forecaster focus due to use of too many interfaces.

- Outmoded customer products due to inability to quickly tailor reports.

6.10.4.3. The RFC modernization project

TVA is addressing this need through the RFC Modernization Project. Throughout 2012, RFC staff partnered with Riverside Technology to evaluate alternatives for a new platform and trial the chosen platform. Delft-FEWS (or simply, FEWS is an open data handling platform initially developed as a hydrological forecasting and warning system) was chosen as the best available alternative and a pilot project was performed to convert the rainfall processing portion of the RFC into FEWS.

Other river management agencies have recently made significant advances developing functional and robust software platforms to support the water management mission. Thus modernization of the RFC System is an entirely realistic, achievable and timely objective for the RFC.

7. ABBRÉVIATIONS

AFC : Contrôle automatique de la fréquence

AVC : Contrôle automatique de la tension

CLDC : Centre de commande de la production hydraulique

CTG : Corporation des Trois Gorges en Chine

ELD : Centre de régulation de la production d'énergie

EDF : Électricité de France

EMS : Management de la production de l'énergie

GIS : Système d'information géographique (SIG)

HPP : Centrale hydroélectrique

ICOLD : Commission Internationale des Grands Barrages (CIGB)

KEPCO : Compagnie Électrique Kansai au Japon

LDCC : Centre local d'exploitation et de surveillance

RDS : Système d'exploitation du réservoir

RFC : Centre de prévision hydraulique

SCADA : Centre de surveillance et d'acquisition des données

TVA : Autorité de contrôle de la vallée du Tennessee

TSO : Opérateur du système électrique

TGP : Projet de la cascade des Trois Gorges

USA : États-Unis d'Amérique

USACE : Corps des Ingénieurs de l'Armée des États-Unis

7.   ABBREVIATIONS

AFC: Automatic Frequency Control
AVC: Automatic Voltage Control
CLDC: Central Load Dispatching Center
CTG: China Three Gorges Corporation
ELD: Economic Load Dispatching
EDF: Electricité de France
EMS: Energy Management System
GIS: Geographic Information System
HPP: Hydropower Plant
ICOLD: International Commission on Large Dams
KEPCO: The Kansai Electric Power Co., Inc.
LDCC: Local Load Dispatching and Control Center
RDS: Reservoir Dispatching System
RFC: River Forecast Centre
SCADA: Supervisory Control and Data Acquisition
TVA: the Tennessee Valley Authority
TSO: Transmission System Operator
TGP: China Three Gorges Project
USA: The United States of America
USACE: the U.S. Army Corps of Engineers

8. REFERENCES

ICOLD Bulletin Preprint-163: Dams for Hydroelectric Energy.

International Hydropower Association, "The Role of Hydropower in Sustainable Development- IHA White Paper", 2003

Xiao Ge. The Basin Climate Characteristics and Prediction Methods in the Key Period of Regulation of Three Gorges Project. Beijing: China Meteorological Press, 2014.

Ma Guangwen. Integrated Operation of Hydropower Stations and Reservoirs. Beijing: China Electric Power Press, 2008.

Claudio Monteiro: Short-term forecasting model for aggregated regional hydropower generation. Energy Conversion and Management, December 2014, Pages 231–238.

Matsuda, S: Practical Application and Continuous Improvement of Rainfall Prediction on Dams for Power Generation in Mountainous Region, the 80th Annual Meeting of ICOLD, June 2012.

Martin Wieland and Rudolf Mueller. Dam safety, emergency action plans and water alarm systems. International Waterpower & Dam Construction, 2009, Pages 34–38.

David L. Bijl, Patrick W. Bogaart, Tom Kram, Bert J.M. de Vries, Detlef P. van Vuuren: Long-term water demand for electricity, industry and households. Environmental Science & Policy, September 2015, Pages 75–86.

Symphorians GR., Madamombe E., van der ZAAG P: Dam Operation for Environmental Water Releases: the Case of Osborne Dam, Save Catchment, Zimbabwe. Physics and Chemistry of the Earth, 2003, Pages 985–993.

Jordan, F.: Real-time flood management by preventive operations on multiple alpine hydropower schemes, in Proceedings of the 31th IAHR Congress, Seoul. IAHR. 2005, Pages 3235–3245.

Frederick N.-F. Chou: Stage-wise optimizing operating rules for flood control in a multi-purpose reservoir. Journal of Hydrology, February 2015, Pages 245–260.

R.U. Kamodkar, D.G. Regulwar: Optimal multiobjective reservoir operation with fuzzy decision variables and resources: A compromise approach. Journal of Hydro-environment Research, December 2014, Pages 428–440.

Yohannes Gebretsadik, Charles Fant, Kenneth Strzepek, Channing Arndt: Optimized reservoir operation model of regional wind and hydro power integration case study: Zambezi basin and South Africa, January 2016, Pages 574–582.

D. Beevers, L. Branchini, V. Orlandini, A. De Pascale, H. Perez-Blanco: Pumped hydro storage plants with improved operational flexibility using constant speed Francis runners. Applied Energy, January 2015, Pages 629–637.

B.R. Mehta, Y.J. Reddy: Chapter 7-SCADA System. Industrial Process Automation Systems, 2015, Pages 237–300.